普通高校"十二五"规划教材

51系列单片机课程设计指导

楼然苗　胡佳文　李光飞
刘玉良　李韵磊　叶继英　编著

北京航空航天大学出版社

内容简介

本书共分 4 个部分：第 1 部分（第 1 章）介绍了单片机课程设计的教学要求；第 2 部分（第 2 章）介绍了一款用于单片机课程设计的实验电路板；第 3 部分（第 3～7 章）介绍了使用 Proteus 仿真软件的 5 个单片机课程设计实例；第 4 部分（第 8～12 章）介绍了 ISD4002 语音录放电路的设计、超声波测距器的设计、简易 LCD 示波器的设计、远程电话控制器的设计、液晶 GPS 定位信息显示器的设计，共 5 个具有代表性的单片机课程设计实例。设计实例均给出了完整的汇编或 C 源程序，为教师教学与学生学习提供了方便。

本书包含的所有设计实例的源程序及 1 个课程设计实验电路板资料，均可在北京航空航天大学出版社网站(www.buaapress.com.cn)的"下载专区"相关页面下载。

本书可作为高等院校单片机课程设计类教材，也可作为电子技术人员的设计参考用书。

图书在版编目(CIP)数据

51 系列单片机课程设计指导 / 楼然苗等编著. -- 北京：北京航空航天大学出版社，2016.2
 ISBN 978 - 7 - 5124 - 2043 - 4

Ⅰ.①5… Ⅱ.①楼… Ⅲ.①单片微型计算机－高等学校－教材 Ⅳ.①TP368.1

中国版本图书馆 CIP 数据核字(2016)第 009358 号

版权所有，侵权必究。

51 系列单片机课程设计指导

楼然苗　胡佳文　李光飞
刘玉良　李韵磊　叶继英　编著

责任编辑　孙兴芳

*

北京航空航天大学出版社出版发行

北京市海淀区学院路 37 号（邮编 100191）　http://www.buaapress.com.cn
发行部电话：(010)82317024　传真：(010)82328026
读者信箱：emsbook@buaacm.com.cn　邮购电话：(010)82316936
艺堂印刷（天津）有限公司印装　各地书店经销

*

开本：710×1 000　1/16　印张：14.25　字数：304 千字
2016 年 2 月第 1 版　2022 年 9 月第 3 次印刷　印数：6 001～7 000 册
ISBN 978 - 7 - 5124 - 2043 - 4　定价：36.00 元

若本书有倒页、脱页、缺页等印装质量问题，请与本社发行部联系调换。联系电话：(010)82317024

前　言

单片机课程设计是学生进一步加深理论知识理解、提高实际单片机设计能力的重要环节,从学生自己设计电路板,到程序编制与调试,最后完成一个单片机系统的设计,可以让学生体验到成功的快乐! 本书内容由4个部分组成:第1部分(第1章)介绍了单片机课程设计的教学要求;第2部分(第2章)介绍了一款用于单片机课程设计的实验电路板;第3部分(第3～7章)介绍了使用Proteus仿真软件的5个单片机课程设计实例;第4部分(第8～12章)介绍了5个具有代表性的单片机课程设计实例。

利用Proteus的单片机硬件电路进行程序运行效果仿真,可以方便、直观地进行单片机程序运行效果演示,极大地拓展了课堂教学及实验教学的硬件环境条件,老师或学生可以在教室或寝室方便地利用计算机进行单片机程序的调试及效果演示,为设计开发单片机应用产品提高效率。为了方便单片机课程设计教学活动的开展,拓展实验场地空间,减少实验材料的消耗,书中第3～7章介绍了DS1302实时时钟、数字温度计、低频信号发生器、16点阵LED显示器、电子密码锁,共5个可采用Proteus软件仿真的课程设计实例,较详细地介绍了其系统功能、设计方案、硬件仿真电路、程序设计、仿真运行结果和参考源程序清单等内容,适合学生在实验室或寝室甚至家里完成单片机课程设计任务,而且用Proteus仿真的单片机程序也可以在真实制作的硬件电路板上运行;书中第8～12章的5个设计实例适合实物制作后再进行程序编程及调试,也可以作为毕业设计的选题内容。本书全部实例源程序及第2章课程设计实验电路板介绍的源程序及电路板制作资料,均可在北京航空航天大学出版社网站(www.buaapress.com.cn)的"下载专区"相关页面下载。

本书第3～7章的Proteus仿真电路图设计采用Proteus 7.1版本,仿真使用时请安装 Proteus 7.1 及以上版本。有关 Proteus 仿真软件的安装与使用方法请参考相关资料,本书不进行相关介绍。

本书可作为高等院校单片机课程设计类教材,也可作为电子技术人员的设计参考用书。

本书在出版、编辑的过程中得到了北京航空航天大学出版社的大力支持,在此表示衷心的感谢;同时对编写中所参考的多部著作的作者表示深深的谢意。

作 者

2016 年 1 月于浙江海洋学院

目　　录

第1章　单片机原理及应用课程设计教学要求 ……………………………… 1
　1.1　单片机原理及应用课程设计教学大纲 ……………………………… 1
　1.2　课程设计教学计划 …………………………………………………… 2
　1.3　系统设计功能要求 …………………………………………………… 4
　1.4　设计报告格式要求 …………………………………………………… 5

第2章　单片机课程设计实验电路板的介绍 ………………………………… 7
　2.1　实验电路板的功能 …………………………………………………… 7
　2.2　实验电路板的电路原理 ……………………………………………… 7
　2.3　实验项目的内容 ……………………………………………………… 10
　2.4　教学实施过程 ………………………………………………………… 11
　2.5　课程设计要求 ………………………………………………………… 11

第3章　DS1302实时时钟的设计 …………………………………………… 12
　3.1　系统功能 ……………………………………………………………… 12
　3.2　设计方案 ……………………………………………………………… 12
　3.3　系统硬件仿真电路的设计 …………………………………………… 12
　3.4　系统程序的设计 ……………………………………………………… 13
　3.5　软件调试与运行结果 ………………………………………………… 14
　3.6　源程序清单 …………………………………………………………… 14
　　3.6.1　汇编源程序清单 ………………………………………………… 14
　　3.6.2　C源程序清单 …………………………………………………… 24

第4章　数字温度计的设计 …………………………………………………… 30
　4.1　系统功能 ……………………………………………………………… 30
　4.2　设计方案 ……………………………………………………………… 30
　4.3　系统硬件仿真电路的设计 …………………………………………… 30
　4.4　系统程序的设计 ……………………………………………………… 31
　4.5　软件调试与运行结果 ………………………………………………… 35
　4.6　源程序清单 …………………………………………………………… 36
　　4.6.1　汇编源程序清单 ………………………………………………… 36
　　4.6.2　C源程序清单 …………………………………………………… 45

第5章　低频信号发生器的设计 ……………………………………………… 50
　5.1　系统功能 ……………………………………………………………… 50

5.2	设计方案	50
5.3	系统硬件仿真电路的设计	50
5.4	系统程序的设计	51
5.5	软件调试与运行结果	53
5.6	源程序清单	54
	5.6.1 汇编源程序清单	54
	5.6.2 C源程序清单	58

第6章 16点阵LED显示器的设计 61

6.1	系统功能	61
6.2	设计方案	61
6.3	系统硬件仿真电路的设计	62
6.4	系统程序的设计	64
6.5	软件调试与运行结果	65
6.6	源程序清单	66
	6.6.1 汇编源程序清单	66
	6.6.2 C源程序清单	73

第7章 电子密码锁的设计 78

7.1	系统功能	78
7.2	设计方案	78
7.3	系统硬件仿真电路的设计	78
7.4	系统程序的设计	81
7.5	软件调试与运行结果	82
7.6	源程序清单	84

第8章 ISD4002语音录放电路的设计 96

8.1	系统功能	96
8.2	设计方案	96
8.3	系统硬件电路的设计	97
8.4	系统程序的设计	99
8.5	调试及性能分析	100
8.6	源程序清单	100

第9章 超声波测距器的设计 108

9.1	系统功能	108
9.2	设计方案	108
9.3	系统硬件电路的设计	109
	9.3.1 单片机系统及显示电路	109
	9.3.2 超声波发射电路	109

9.3.3 超声波检测接收电路 …………………………………………… 109
9.4 系统程序的设计 ……………………………………………………… 111
    9.4.1 超声波测距器的算法设计 …………………………………… 111
    9.4.2 主程序 ………………………………………………………… 112
    9.4.3 超声波发生子程序和超声波接收中断程序 ………………… 112
9.5 调试及性能分析 ……………………………………………………… 113
    9.5.1 调试 …………………………………………………………… 113
    9.5.2 性能指标 ……………………………………………………… 113
9.6 源程序清单 …………………………………………………………… 113
    9.6.1 汇编源程序清单 ……………………………………………… 113
    9.6.2 C源程序清单 ………………………………………………… 120

第10章 简易LCD示波器的设计 ……………………………………… 124
10.1 系统功能 …………………………………………………………… 124
10.2 设计方案 …………………………………………………………… 124
10.3 系统硬件电路的设计 ……………………………………………… 124
10.4 系统程序的设计 …………………………………………………… 127
10.5 调试及性能分析 …………………………………………………… 129
10.6 源程序清单 ………………………………………………………… 129

第11章 远程电话控制器的设计 ………………………………………… 141
11.1 系统功能 …………………………………………………………… 141
11.2 设计方案 …………………………………………………………… 141
11.3 系统硬件电路的设计 ……………………………………………… 142
    11.3.1 电话振铃检测电路 ………………………………………… 142
    11.3.2 自动模拟摘机电路 ………………………………………… 142
    11.3.3 DTMF信号解码电路 ……………………………………… 142
    11.3.4 语音提示电路 ……………………………………………… 145
    11.3.5 单片机控制电路 …………………………………………… 146
    11.3.6 接口驱动电路 ……………………………………………… 147
    11.3.7 电源电路 …………………………………………………… 148
11.4 系统程序的设计 …………………………………………………… 148
    11.4.1 语音录音程序 ……………………………………………… 148
    11.4.2 语音放音程序 ……………………………………………… 149
    11.4.3 双音频解码数据读码程序 ………………………………… 149
    11.4.4 自动摘机处理程序 ………………………………………… 150
11.5 调试及性能分析 …………………………………………………… 151
11.6 源程序清单 ………………………………………………………… 151

第 12 章 液晶 GPS 定位信息显示器的设计 …… 162

12.1 系统功能 …… 162
12.2 设计方案 …… 162
12.2.1 GPS 模块的选择 …… 162
12.2.2 显示器的选择 …… 163
12.2.3 CPU 的选择 …… 164
12.3 系统硬件电路的设计 …… 164
12.3.1 电源电路 …… 166
12.3.2 AT89C52 单片机系统 …… 166
12.3.3 键盘电路 …… 167
12.3.4 单片机与 GPS-OEM 板接口电路 …… 168
12.3.5 单片机与液晶显示器接口电路 …… 172
12.4 系统程序的设计 …… 173
12.4.1 系统软件设计原理 …… 173
12.4.2 LCD 液晶显示器程序 …… 174
12.4.3 GPS 接收子程序 …… 182
12.4.4 键盘子程序 …… 183
12.4.5 显示子程序 …… 183
12.4.6 初始化子程序 …… 186
12.4.7 主程序 …… 187
12.5 调试及性能分析 …… 188
12.5.1 调试步骤 …… 188
12.5.2 性能分析 …… 188
12.6 源程序清单 …… 188

附录 A 80C51 系列单片机的特殊功能寄存器表 …… 210
附录 B 80C51 系列单片机中断入口地址表 …… 212
附录 C 80C51 系列单片机汇编指令表 …… 213
参考文献 …… 219

第 1 章　单片机原理及应用课程设计教学要求

1.1　单片机原理及应用课程设计教学大纲

课程名称(中文)： 单片机原理及应用课程设计
课程名称(英文)： Course Design about Principles & Applications of Microcontroller
课程编码： 0433240
学时/学分： 1.5 周/1.5
开课学期： 第 5 学期
课程性质： 集中实践教学
适用专业： 电子信息工程、电气工程及自动化、船舶电子电气工程

1. 课程设计的教学目的与任务

本课程设计要求学生在 1.5 周内自选一个单片机应用系统设计，并完成设计报告。通过设计实践，使学生能够掌握单片机的应用特点、汇编或 C 语言的编程方法，学会规范书写单片机应用系统设计的报告，为毕业设计打下良好的基础。

2. 课程设计的教学内容与基本要求

(1) 课程设计的题目
课程设计的题目可从以下几个方面参考选择：
- 彩灯控制器设计；
- 单片机时钟设计；
- DS1302 实时时钟设计；
- 数字温度计设计；
- 低频信号发生器设计；
- 16 点阵 LED 显示器设计；
- 单片机直流电压表的设计；
- 单片机的鼠标操作控制设计；
- 单片机电子密码锁的设计；
- 单片机在其他领域的综合应用设计(学生自拟题)。

(2) 具体要求
① 用 Proteus 软件仿真的，应完成硬件及软件的设计，并能演示系统功能。利用

实验室现成实验电路板或自行设计实物进行程序功能设计的,要能演示程序实现功能。

② 完成设计报告,设计报告的内容及格式由教师提供。

3. 课程设计的方法

课程设计可以在实验室集中或分散进行,电路板硬件由实验室提供,设计报告用电子稿打印一份上交并同时上交所有电子设计文档,每个学生均独立选题并在教师的指导下完成设计任务。

4. 与相关课程的联系

先修课程:计算机文化基础、C语言程序设计、模拟电子技术、数字电子技术、单片机原理及应用、电路原理图与PCB设计等。

5. 时间与学时分配

① 课程设计的学习时间宜安排在期末或短学期。

② 学时分配:共1.5周(需要答辩的时间(含答辩)1~2天)。

6. 考核与成绩评定

课程设计的成绩根据学生设计实物、设计报告,并结合程序功能演示与答辩情况,按五级评分制(优、良、中、及格、不及格)进行综合评定。

7. 教材与教学参考书

[1] 楼然苗,胡佳文,李光飞,等. 51系列单片机课程设计指导[M]. 北京:北京航空航天大学出版社,2015.

[2] 楼然苗,胡佳文,李光飞,等. 51系列单片机原理及应用[M]. 北京:北京航空航天大学出版社,2014.

8. 说　明

课程设计在实验室场地、时间不允许的情况下可分散进行,但必须组织答辩来评定成绩。

1.2　课程设计教学计划

1. 课程设计的对象

课程设计的对象包括电气工程及自动化、电子信息工程、船舶电子电气工程等专业的学生。

2. 课程设计的任务和目的

本课程设计要求学生在1.5~2周内编程设计一个单片机应用系统,并完成设计报告。通过设计实践,使学生能够掌握单片机的应用特点、编程方法,学会规范书写单片机实际应用系统的设计开发过程及设计的报告,为毕业设计打下良好的基础。

3. 课程设计的内容及要求

（1）课程设计的题目

课程设计的题目可从以下几个方面参考选择：

① 单片机在计时控制方面的应用设计，如：时钟、频率计、彩灯、交通灯。

② 单片机在计数控制方面的应用设计，如：计数器、计分器、抢答器、报警器。

③ 单片机在运算控制方面的应用设计，如：密码锁、计算器、乒乓球游戏机。

④ 单片机在波形发生方面的应用设计，如：简易电子琴、音乐盒、LED 调光灯。

⑤ 单片机在通信技术方面的应用设计，如：双机通信、PC 可控单片机系统、对话机器人。

⑥ 单片机 A/D 转换技术方面的应用设计，如：电压计、温度计、照度计。

⑦ 自选单片机其他应用设计（选题后必须经老师批准）。

（2）具体要求

① 完成控制实物设计（或电路仿真）及程序的编制，能演示系统功能。

② 完成设计后上交纸质设计报告 1 份及所有电子设计文档。

③ 系统功能要求参照 1.3 节的内容或学生自拟，设计报告格式规范见 1.4 节。

④ 完成作品功能演示及答辩。

4. 时间与学时安排

① 课程设计时间安排在本学期第××周～第××周。

② 总体教学时间安排如表 1.1 所列。

表 1.1　总体教学时间安排

教学任务与内容	时　　间	辅导教师	地　点
课程设计任务布置与学生选题	2 学时	×××电话：	实验楼或教室
课程设计辅导	规定学时内学生在实验室或寝室用实验电路板完成设计任务	×××电话：	实验楼
课程设计报告上交与答辩	在规定时间前由班长收齐后统一交老师办公室	×××电话：	实验楼或教室

5. 考核与成绩评定

课程设计成绩根据学生设计报告并结合程序功能演示与答辩的情况，按优、良、中、及格、不及格五级评分制进行综合评定。

6. 评分标准

① 程序功能不能演示、设计工作量 70% 由别人代做或抄袭的、缺课时间达 1/3 的，符合以上其中任意一条的，成绩评定为不及格。

② 能实现简单的应用程序（课程设计题目①类）功能演示，设计报告的版面格式、文字描述的观点及语法、图表程序基本正确的，成绩评定为及格。

③ 能实现一般的应用程序(课程设计题目②类)功能演示,设计报告的版面格式、文字描述的观点及语法、图表程序基本正确的,成绩评定为中等。

④ 能实现较复杂的应用程序(课程设计题目③、④类)功能演示,设计报告的版面格式、文字描述的观点及语法、图表程序较正确的,成绩评定为良好。

⑤ 能实现复杂的应用程序(课程设计题目⑤、⑥类)功能演示,设计报告的版面格式、文字描述的观点及语法、图表程序较正确的,成绩评定为优秀。

⑥ 在校期间,将设计报告以论文形式在正式杂志上发表,或将设计内容申请并取得实用或发明专利证书的,本课程成绩最后均认定为优秀。

1.3 系统设计功能要求

单片机原理及应用课程设计的功能要求可以参考以下几个方面或学生自拟。

(1) 单片机在计时控制方面应用的设计功能要求

① 时钟:能计时,可校准时间,至少有一种附带功能(如秒表、定时器或闹钟功能)。

② 频率计:能测试并显示 1 Hz~10 kHz 频率、5 V 的方波,可附带方波发生器功能。

③ 彩灯:要求控制 16 个 LED 有两种以上的闪烁方式。

④ 交通灯:要求模拟控制十字路口交通信号,有倒计时显示。

(2) 单片机在计数控制方面应用的设计功能要求

① 计数器:利用外中断通过对传感器送来的方波计数的方法,实现对产品线上的产品计数,用数码管显示计数结果。

② 计分器:设计一个用于竞技比赛的记分牌,通过按钮设定对双方比分的加减,可附加定时功能。

③ 抢答器:要求有 4 路以上,可以设定限时及其他功能。

④ 报警器:要求能对 8 路以上的通道进行巡检,并能够进行声光报警和显示通道号。

(3) 单片机在运算控制方面应用的设计功能要求

① 密码锁:要求可以通过键盘设定 6 位以上的密码,密码正确才开锁,设置时可显示密码,开锁时不显示密码。

② 计算器:要求能带小数进行加、减、乘、除计算。

③ 乒乓球游戏机:设计一个用 LED 灯模拟乒乓球运动过程的游戏,以按键代表球拍,以亮着的小灯代表乒乓球,可附带记分功能。

(4) 单片机在波形发生方面应用的设计功能要求

① 简易电子琴:可以显示音符,能弹奏简单乐曲。

② 音乐盒:能播放两首以上乐曲,可以显示乐曲编号。

③ LED 调光灯:单片机利用 PWM 原理对 8 个 LED 进行 4 级亮度调节,采用按钮方式进行调节。

(5) 单片机在通信技术方面应用的设计功能要求

① 双机通信:设计制作一个两个单片机系统相互通信的模型,使用 UART 或 SPI,可利用按钮进行操作,数码管显示。

② PC 可控单片机系统:通过计算机键盘或鼠标操作,至少可以发送 8 种以上控制命令,使得单片机执行并进行数码管显示。

③ 对话机器人:可以通过计算机终端与单片机进行文字对话,要求 10 句以上,并基本符合逻辑。

(6) 单片机 A/D 转换技术方面应用的设计功能要求

① 电压计:能够同时测量 1~8 路的直流电压,并且能够轮流显示读数值,测量量程在 3 V 以内。

② 温度计:用数码管能够显示实时的室内温度,要求能够设定报警温度。

③ 照度计:用数码管能够显示实时的室内亮度,要求能够设定报警照度。

(7) 自选题目

控制功能由学生自己确定。

1.4　设计报告格式要求

设计报告题目(三号宋体加粗居中)

姓名、班级、学号(小四号宋体居中)

1. 系统功能的确定(小四号宋体加粗)

正文(小四号宋体)

2. 方案论证

2.1　方案一

2.2　方案二

2.3　方案三

选定系统方案并给出总体框图。

3. 系统硬件的设计

3.1　主控制器的设计(电路图及设计说明)

3.2　接口电路的设计

⋮

4. 系统软件的设计

4.1　主程序的设计(程序流程图及说明)

4.2　按键扫描程序的设计

⋮

5. 系统调试

5.1　硬件调试

5.2　软件调试

5.3　综合调试

6. 指标测试

6.1　测试仪器

6.2　指标测试

7. 结　论

对课程设计的结果进行总结。

参考文献：

[1] 楼然苗,胡佳文,李光飞,等.51系列单片机原理及设计实例[M].北京:北京航空航天大学出版社,2010.

[2] 楼然苗,胡佳文,李光飞,等.单片机实验与课程设计(Proteus仿真版)[M].杭州:浙江大学出版社,2010.

[3] 楼然苗,李光飞.单片机课程设计指导[M].2版.北京:北京航空航天大学出版社,2012.

[4] 楼然苗,胡佳文,李光飞,等.单片机实验与课程设计指导(Proteus仿真版)[M].2版.杭州:浙江大学出版社,2013.

[5] 楼然苗,胡佳文,李光飞,等.51系列单片机原理及应用[M].北京:北京航空航天大学出版社,2014.

第 2 章　单片机课程设计实验电路板的介绍

2.1　实验电路板的功能

"单片机原理及应用"作为工科类大学生的一门专业基础课,具有实用性强、难学的特点,因此,单片机课程的实验内容对学生实践动手创造能力的培养尤为重要。本章将介绍一套用于单片机课程设计实验的电路板设计方案,其采用中文液晶显示器,教师及学生在板上可做实时时钟编程实验、数显温度计编程实验、超声波测距编程实验、红外线遥控发射与接收编程实验、正弦波信号源编程实验、串行通信编程实验、PS/2 鼠标实验、音乐编程实验和数据存储器实验等,并且在编程实验中通过将以上实验项目进行适当组合,即可成为功能复杂的单片机应用设计项目。另外,上述编程实验非常适合在学习单片机理论知识后进行单片机应用的综合设计训练。实验过程可从学生焊接元件开始到编程实现控制功能,这既培养了学生的学习兴趣,又使学生掌握了单片机设计的方法与过程,教学效果很好。

2.2　实验电路板的电路原理

图 2.1 所示为单片机综合实验电路板的电路设计原理图,由单片机控制器、中文液晶显示器、实时时钟、测温传感器、红外线发射及接收电路、I^2C 存储器、超声波发射及接收电路、D/A 转换器、PS/2(3D 鼠标)接口、旋转编码开关、串行通信接口、LED 小灯、蜂鸣器、按键开关以及电源电路等组成。

1. 单片机控制器

单片机采用宏晶公司生产的 STC 系列单片机,可随时修改程序并可以从串口或 USB 口下载到实验电路板中实时运行程序,下载工具可从宏晶公司的网站上获取。实验中常用的单片机型号为 STC12C5A16S2,单片机的运行速度是普通单片机的 8～12 倍,并且内带 ADC、PWM、存储、片内时钟等资源,特别适用于学生实验及开发设计选用。

2. 中文液晶显示器

中文液晶显示器采用 12232F,它是一种内置 8 192 个 16×16 点中文汉字库和 128 个 16×8 点 ASCII 字符集图形点阵液晶显示器。它可完成图形显示,也可以显示 7.5 个×2 行(16×16 点阵)中文汉字。该显示器与单片机接口可采用并行或串行方式控制,实验电路中采用串行控制方式,可进行背光灯的控制及中英文与数字的显示刷新。

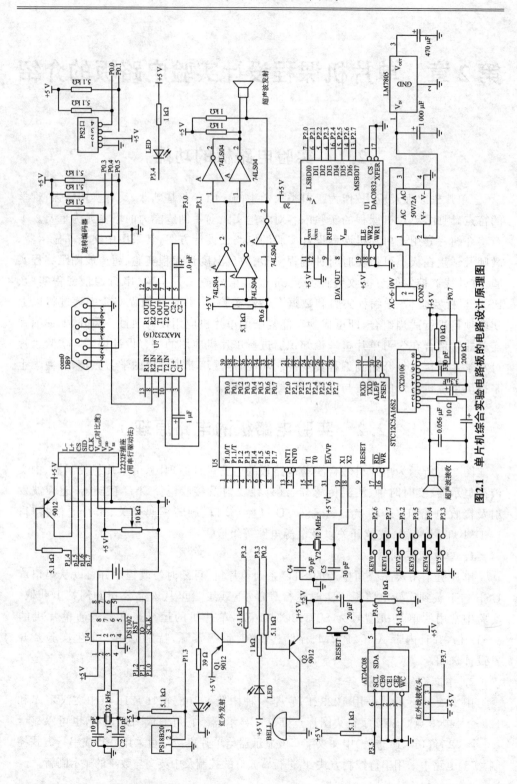

图2.1 单片机综合实验电路板的电路设计原理图

3. 实时时钟

实时时钟芯片使用美国 DALLAS 公司的 DS1302,它是一种高性能、低功耗、带 RAM 的实时计时芯片。DS1302 不仅计时准确,而且可以在电流很小的后备电源（2.5～5.5 V 电源,在 2.5 V 时耗电小于 300 nA）下继续计时。DS1302 时钟芯片包括实时时钟/日历和 31 个字节的静态 RAM,它经过一个简单的串行接口与单片机通信。实时时钟提供秒、分、时、日、周、月和年等信息,对于小于 31 天的月和月末的日期会自动进行调整,此外,还包括闰年校正的功能;时钟的运行可以采用 24 小时或带 AM(上午)/PM(下午)的 12 小时格式。它采用三线接口与单片机进行同步通信。DS1302 有主电源/后备电源双电源引脚,V_{CC1} 在单电源与电池供电的系统中提供低电源并提供低功率的电池备份,V_{CC2} 在双电源系统中提供主电源。

4. 测温传感器

测温传感器采用美国 DALLAS 半导体公司继 DS1820 之后推出的一种改进型智能温度传感器 DS18B20。DS18B20 的测温范围为 −55～125 ℃,分辨率最大可达 0.062 5 ℃。其可以根据实际要求通过简单的编程实现 9～12 位的数字值读数,采用单线与单片机通信,减少了外部的硬件电路,具有低成本和易使用的特点。

5. 红外线发射及接收电路

红外线遥控信息码由单片机的定时器调制成 38.5 kHz 的红外线载波信号,通过三极管 9012 放大后由红外线发射管发送。红外线接收处理采用通用的集成模块化三引脚红外接收器,输出为检波整形过的方波信号。

6. I²C 存储器

存储器采用 AT24C08,采用 I²C 通信,仅需要 2 条 I/O 口就可以进行存储操作。

7. 超声波发射及接收电路

超声波发射电路主要由缓冲反向器 74LS04 和超声波换能器构成。单片机端口输出的 38.5 kHz 方波信号一路经一级反向器后送到超声波换能器的一个电极,另一路经两级反向器后送到超声波换能器的另一个电极,用这种推挽形式将方波信号加到超声波换能器两端就可以提高超声波的发射强度。输出端采用两个反向器并联,用以提高驱动能力;上拉电阻一方面可以提高反向器 74LS04 输出高电平的驱动能力,另一方面可以增加超声换能器的阻尼效果,缩短其自由振荡的时间。

超声波接收由 CX20106 完成,CX20106 内部具有前置放大、载波选频、脉冲解调等功能。在收到超声波时,CX20106 的引脚 7 输出低电平。

8. D/A 转换器

DAC0832 是 8 位 D/A 转换器,属于 8 位电流输出型 D/A 转换器,转换时间为 1 μs,片内带输入数字锁存器。DAC0832 与单片机接成数据直接写入方式,当单片机把一个数据直接写入 DAC 寄存器时,DAC0832 的输出模拟电压信号将随之对应变化。利用 D/A 转换器可以产生各种波形,如方波、三角波、锯齿波等,以及由它们组合产生的复合波形和不规则波形,这些复合波形利用标准的测试设备是很难产生的。

9. PS/2(3D 鼠标)接口

在 PS/2 接口中可用 PC 键盘和鼠标进行与单片机的通信实验。PS/2 接口只要占用 2 根端口线即可实现对单片机应用系统的控制,在有液晶或 CRT 显示器的系统中使用将会非常方便,是嵌入式设计中人机接口的首选设计方案。实验中使用 3D 鼠标的 3 个按键可控制 2 个小灯的亮灭和 1 个蜂鸣器的鸣叫,鼠标在平面上的移动可以在液晶显示器上观察到相应水平方向、竖直方向数据的变化,同时滚轮的前后拨动也能在液晶显示器上看到相应数据的增减。

10. 旋转编码开关

旋转编码开关可产生正转和反转的脉冲个数数据及一个按压开关信号,这在需要调节参数变量的程序中应用非常方便。

11. 串行通信接口

串行口通信电路可与 PC 超级终端进行通信实验,也是程序下载时必需的。采用宏晶公司的 STC 系列,可以在该公司的网站上下载免费软件,学生编程时可随时进行程序的下载运行。

12. LED 小灯及蜂鸣器

实验电路板上设计了 2 个 LED 小灯及 1 个蜂鸣器用于状态指示。

13. 电源电路

电源电路可在交直流输入下工作,稳压电路采用 LM7805 稳压集成芯片。

2.3 实验项目的内容

实验电路板可进行 8 个以上的综合性课程设计编程实验项目,具体如下。

① 实时时钟编程实验:可在液晶屏上显示年、月、日、星期、时、分、秒等信息,可进行实时时间的调整,可设定多次定时功能,能在 EEPROM 中存储定时数据,能编写按键音功能,能编写整点报时功能等。

② 数显温度计编程实验:能在液晶屏上显示当前的气温、水温或其他被测物的温度,能设定低温或高温报警,能模拟空调等温控器的作用,能在 EEPROM 中存储设定的报警温度。

③ 超声波测距编程实验:能显示障碍物的距离,能设定远距离或近距离的报警,能模拟运动物体的自动测距报警功能,能在 EEPROM 中存储距离报警数据。

④ 遥控发射与接收编程实验:能用一块实验板作为遥控器,另一块作为接收器进行红外线遥控编程实验。遥控器有 6 个按键,接收器的功能演示可用 LED 小灯、蜂鸣器、液晶背光灯等。

⑤ 正弦波信号源编程实验:可输出 0.01~83 Hz 的正弦波(或三角波),可输出 1.3 Hz~10.6 kHz 的方波信号。

⑥ 串行通信编程实验:能与 PC 进行串行通信,在 PC 的超级终端上显示中文字

符或其他字符;能用 PC 发命令控制单片机的功能操作。

⑦ 音乐编程实验:能用蜂鸣器演奏自编歌曲。

⑧ PS/2 鼠标实验:能用鼠标的 3 个按键操作实验电路板上的小灯及蜂鸣器,鼠标在平面上的移动可以在液晶显示器上看到相应水平方向、竖直方向数据的变化,同时滚轮的前后拨动也能在液晶显示器上看到相应数据的增减。

⑨ 将以上实验项目进行组合并结合鼠标、旋转编码开关等,使其成为多功能的应用设计实验项目。

2.4 教学实施过程

单片机综合性课程设计实验一般安排在学期末训练,在做设计实验之前,必须提前将实验设计项目的设计原理及编程思想在课堂教学中进行介绍,结合学生平时进行的上机小实验,在学生有一定编程基础的条件下才能进行课程设计实验。参考课程设计学时安排为:焊接 4 学时(半天),编程调试及设计报告 1 周(5 天),答辩 1 天。课程设计实施过程如下:

① 按实验材料清单领取元件并焊接实验电路板。

② 检查硬件焊接的正确性,测试电路板。

③ 编制调试程序,然后下载、脱机运行试验。

④ 程序调试完成后编写设计报告。

⑤ 答辩评定成绩。

2.5 课程设计要求

设计前学生应查阅单片机、12232F 液晶显示器、DS1302 实时时钟、AT24C08 存储器、RS2 通信串口芯片 MAX232、DAC0832 数/模转换芯片、CX20106 红外线接收芯片、74LS04(六反相器)芯片、DS18B20 数字测温芯片等的资料,了解其使用特性,阅读实验基本演示程序,然后领取元件、焊接、编制调试程序、烧录程序、脱机运行试验,最后完成设计报告。设计要求系统开机时能显示校名、班级、学号、姓名等信息,在温度计、超声测距器、时钟计时器等实验项目中选一个,并结合自己的能力设计完善各种程序控制功能。

单片机设计实验内容的趣味性是提高学生学习兴趣的重要条件,单片机课程设计实验电路板成本为 70 元左右,远远低于市场上实验箱的价格,而且每年可对个别实验项目进行调整修改,是单片机课程设计实验教学中较为理想的选择方案。本书配套资料中有本章实验电路板的电子线路图及 PCB 资料,可作为教师或学生课程设计实验时的参考。更多资料详见浙江海洋学院单片机原理及应用精品课程网站 http://61.153.216.116/jpkc/jpkc/dpj/。

第 3 章 DS1302 实时时钟的设计

3.1 系统功能

DS1302 实时时钟芯片能输出阳历年、月、日,以及星期、小时、分、秒等计时信息,可制作成实时时钟。本系统要求用 8 位 LED 数码管实时显示时、分、秒时间。

3.2 设计方案

按照系统设计功能的要求,确定由主控模块、时钟模块、显示模块、键盘接口模块、发声模块共 5 个模块组成,电路系统构成框图如图 3.1 所示。主控芯片使用 AT89C52 单片机,时钟芯片使用美国 DALLAS 公司推出的一种高性能、低功耗、带 RAM 的实时时钟 DS1302。采用 DS1302 作为计时芯片,可以做到计时准确,更重要的是,DS1302 可以在电流很小的后备电源(2.5~5.5 V 电源,在 2.5 V 时耗电小于 300 nA)下继续计时,而且 DS1302 可以编程选择多种充电电流来对后备电源进行慢速充电,可以保证后备电源基本不耗电。显示电路采用 8 位共阳 LED 数码管,采用查询法查键实现功能调整。

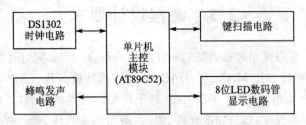

图 3.1 DS1302 实时时钟电路系统构成框图

3.3 系统硬件仿真电路的设计

DS1302 实时时钟的 Proteus 硬件仿真电路如图 3.2 所示。时钟芯片的晶振频率为 32.768 kHz,3 个数据、时钟、片选口可不接上拉电阻;LED 数码管采用动态扫描方式显示,P0 为段码输出口,P2 口为扫描驱动口,扫描驱动信号经 74HC244 功率放大用作 LED 点亮电源;调时按键设计了 2 个,分别接在 P3.5 口和 P3.6 口,用于设定及加 1 调整;P3.7 口连接了一个蜂鸣器,用于按键发声提醒。

第 3 章　DS1302 实时时钟的设计

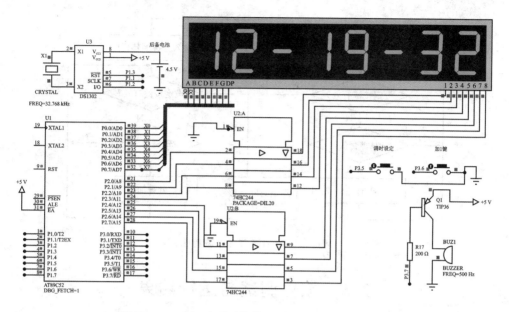

图 3.2　DS1302 实时时钟的 Proteus 硬件仿真电路图

3.4　系统程序的设计

1. 时钟读出程序的设计

因为使用了时钟芯片 DS1302,所以时钟程序只须从 DS1302 各个寄存器中读出年、月、日、周、时、分、秒等数据再处理即可,本次设计中仅读出时、分、秒数据。在首次对 DS1302 进行操作之前,必须对它进行初始化,然后从 DS1302 中读出数据,再经过处理后送给显示缓冲单元。时钟读出程序流程图见图 3.3。

2. 时间调整程序的设计

调整时间用两个调整按钮,一个作为设定控制用,另一个作为加 1 调整用。在调整时间的过程中,要调整的那位应与别的位有所区别,所以增加了闪烁功能,即调整的那位一直在闪烁,直到调整下一位。闪烁原理就是使要调整的那位,每隔一定时间熄灭一次,比如说 50 ms,利用定时器计时,当达到

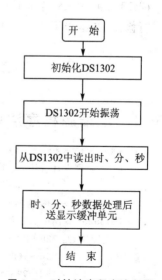

图 3.3　时钟读出程序流程图

50 ms 时,就送给该位熄灭符,在下一次溢出时,再送正常显示的值,不断交替,直到调整该位结束。时间调整程序流程图如图 3.4 所示。

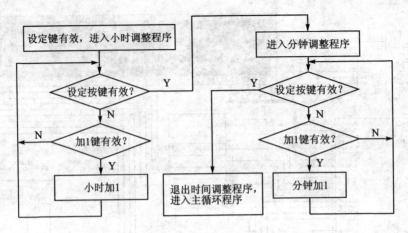

图 3.4 时间调整程序流程图

3.5 软件调试与运行结果

调试分为 Proteus 硬件电路调试和程序软件调试:硬件电路调试主要是检查各元件的连接线是否接好,另外,可以通过编写一个小的调试软件来测试硬件电路是否正常;软件调试应分块进行,先进行显示程序调试,再编写 DS1302 芯片的读/写程序,最后通过多次修改与完善达到理想的功能效果。

DS1302 的晶振频率是计时精度的关键,在实际设计中可换用标准晶振或用小电容进行修正,在本仿真电路中不需要对计时精度进行校准。

3.6 源程序清单

3.6.1 汇编源程序清单

```
;*****************************************************;
;           课程设计程序:DS1302 实时时钟                ;
;                  12 MHz 晶振                         ;
;*****************************************************;
; 从 DS1302 中读出的数据放在 52H(小时)51H(分钟)50H(秒)
; 显示缓冲单元:72H~73H(45H~44H)小时
;              75H~76H(43H~42H)分钟
;              78H~79H(41H~40H)秒
; 定时器 T1 为调整时闪烁用
; 显示式样:15 - 38 - 12
;****************定义*********************
;
```

第 3 章 DS1302 实时时钟的设计

```
            SCLK        EQU  P1.1     ;   DS1302 时钟口,DS1302 第 7 引脚
            IO          EQU  P1.2     ;   数据口,DS1302 第 6 引脚
            RST         EQU  P1.3     ;   使能口,DS1302 第 5 引脚
            KEYSW0      EQU  P3.5     ;   调时按键
            KEYSW1      EQU  P3.6     ;   加 1 按键
            BELL        EQU  P3.7
            hour        DATA 52H      ;   DS1302 读出时
            minute      DATA 51H      ;   DS1302 读出分
            second      DATA 50H      ;   DS1302 读出秒
            DS1302_ADDR DATA 32H      ;   DS1302 需操作的地址数据存放
            DS1302_DATA DATA 31H      ;   DS1302 读出或需写入的数据存放
            INTCON      DATA 30H      ;   闪烁中断计时用
            CON_DATA    DATA 06H      ;   闪烁时间 = 65 × 6 = 0.39 s
            DISPFIRST   EQU  33H      ;   显示地址首址
            DELAYR3     EQU  38H      ;   延时程序使用的寄存器
            DELAYR5     EQU  39H
            DELAYR6     EQU  3AH
            DELAYR7     EQU  3BH
;
;***********************程序入口*********************;
;
            ORG    0000H
            LJMP   START
            ORG    0003H
            RETI
            ORG    000BH
            RETI
            ORG    0013H
            RETI
            ORG    001BH
            LJMP   INTT1
            ORG    0023H
            RETI
            ORG    002BH
            RETI
;
;*******************主程序********************;
;
START:
            MOV    SP,#80H           ;堆栈在 80H 上
            CLR    RST               ;DS1302 禁止
            MOV    DISPFIRST,#72H
```

```
        MOV     74H,#12                 ;"-"
        MOV     77H,#12                 ;"-"
        MOV     TMOD,#10H               ;计数器1,方式1
        MOV     TL1,#00H
        MOV     TH1,#00H
        MOV     INTCON,#CON_DATA
        CLR     00H                     ;清闪烁标志
        CLR     01H                     ;清闪烁标志
        SETB    EA
        MOV     DS1302_ADDR,#8EH
        MOV     DS1302_DATA,#00H        ;允许写DS1302;#80,禁止
        LCALL   WRITE
        MOV     DS1302_ADDR,#90H
        MOV     DS1302_DATA,#0A6H
;DS1302充电电流为1.1 mA;#A5为2.2 mA;#A7为0.6 mA
        LCALL   WRITE
        MOV     DS1302_ADDR,#80H
        MOV     DS1302_DATA,#00H        ;DS1302晶振开始振荡;#80H,禁止
        LCALL   WRITE
;
;以下为主程序
MAIN1:  MOV     DS1302_ADDR,#85H        ;读出小时
        LCALL   READ
        MOV     hour,DS1302_DATA
        LCALL   DISPLAY                 ;显示刷新
        MOV     DS1302_ADDR,#83H        ;读出分钟
        LCALL   READ
        MOV     minute,DS1302_DATA
        LCALL   DISPLAY                 ;显示刷新
        MOV     DS1302_ADDR,#81H        ;读出秒
        LCALL   READ
        MOV     second,DS1302_DATA
        LCALL   DISPLAY                 ;显示刷新
        MOV     R0,hour                 ;小时分离,送显示缓冲单元缓存
        LCALL   DIVIDE
        MOV     73H,R1                  ;时个位
        MOV     44H,R1
        MOV     72H,R2                  ;时十位
        MOV     45H,R2
        LCALL   DISPLAY                 ;显示刷新
        MOV     R0,minute               ;分钟分离,送显示缓冲单元缓存
        LCALL   DIVIDE
```

第 3 章 DS1302 实时时钟的设计

```
            MOV     76H,R1              ;分个位
            MOV     42H,R1
            MOV     75H,R2              ;分十位
            MOV     43H,R2
            LCALL   DISPLAY             ;显示刷新
            MOV     R0,second           ;秒分离,送显示缓冲单元缓存
            LCALL   DIVIDE
            MOV     79H,R1              ;秒个位
            MOV     40H,R1
            MOV     78H,R2              ;秒十位
            MOV     41H,R2
            LCALL   DISPLAY             ;显示刷新
;
            JNB     KEYSW0,SETG         ;调整时间控制键
            LJMP    MAIN1
;
;****************** 公历设置程序 ********************;
;
        SETG:
            LCALL   DL20MS
            JB      KEYSW0,MAIN1
WAITKEY0:   LCALL   DISPLAY             ;等待按键释放
            JNB     KEYSW0,WAITKEY0
            LCALL   DISPLAY
            JNB     KEYSW0,WAITKEY0
            MOV     78H,#00H            ;调时时秒十位数为 0
            MOV     79H,#00H            ;调时时秒个位数为 0
            MOV     40H,#00H            ;调时时秒十位数为 0
            MOV     41H,#00H            ;调时时秒个位数为 0
            MOV     DS1302_ADDR,#8EH
            MOV     DS1302_DATA,#00H    ;允许写 DS1302
            LCALL   WRITE
            MOV     DS1302_ADDR,#80H
            MOV     DS1302_DATA,#80H    ;DS1302 停止振荡
            LCALL   WRITE
            SETB    TR1                 ;闪烁开始
            SETB    ET1
;
        SETG9:  LCALL   DISPLAY         ;等待键按下
            JNB     KEYSW0,SETG10
            JNB     KEYSW1,GADDHOUR
            AJMP    SETG9
```

```
           GADDHOUR: LCALL   DL20MS
                    JB      KEYSW1,SETG9
                    MOV     R7,52H              ;小时加1
                    LCALL   ADD1
                    MOV     52H,A
                    CJNE    A,#24H,GADDHOUR11
         GADDHOUR11: JC     GADDHOUR1
                    MOV     52H,#00H
          GADDHOUR1: MOV    DS1302_ADDR,#84H    ;小时值送入 DS1302
                    MOV     DS1302_DATA,52H
                    LCALL   WRITE
                    MOV     R0,52H
                    LCALL   DIVIDE              ;小时值分离送显示缓冲单元缓存
                    MOV     73H,R1
                    MOV     44H,R1
                    MOV     72H,R2
                    MOV     45H,R2
           WAITKEY1: LCALL  DISPLAY             ;等待按键释放
                    JNB     KEYSW1,WAITKEY1
                    LCALL   DISPLAY
                    JNB     KEYSW1,WAITKEY1
                    AJMP    SETG9
;
            SETG10: LCALL   DL20MS
                    JB      KEYSW0,SETG9
                    SETB    01H                 ;调分时候的闪烁标志
          WAITKEY00: LCALL  DISPLAY             ;等待按键释放
                    JNB     KEYSW0,WAITKEY00
                    LCALL   DISPLAY
                    JNB     KEYSW0,WAITKEY00
            SETG11: LCALL   DISPLAY             ;等待分调整
                    JNB     KEYSW0,SETGOUT
                    JNB     KEYSW1,GADDMINTUE
                    AJMP    SETG11
;
         GADDMINTUE: LCALL  DL20MS
                    JB      KEYSW1,SETG11
                    MOV     R7,51H              ;分钟加1
                    LCALL   ADD1
                    MOV     51H,A
                    CJNE    A,#60H,GADDMINTUE11
        GADDMINTUE11: JC    GADDMINTUE1
```

第 3 章 DS1302 实时时钟的设计

```
              MOV     51H,#00H
GADDMINTUE1:  MOV     DS1302_ADDR,#82H    ;分钟值送入 DS1302
              MOV     DS1302_DATA,51H
              LCALL   WRITE
              MOV     R0,51H
              LCALL   DIVIDE              ;分钟值分离送显示缓冲单元缓存
              MOV     76H,R1
              MOV     42H,R1
              MOV     75H,R2
              MOV     43H,R2
WAITKEY111:   LCALL   DISPLAY             ;等待按键释放
              JNB     KEYSW1,WAITKEY111
              LCALL   DISPLAY
              JNB     KEYSW1,WAITKEY111
              AJMP    SETG11


SETGOUT:      LCALL   DL20MS
              JB      KEYSW0,SETG11
              MOV     DS1302_ADDR,#80H
              MOV     DS1302_DATA,#00H    ;DS1302 晶振开始振荡
              LCALL   WRITE
              MOV     DS1302_ADDR,#8EH
              MOV     DS1302_DATA,#80H    ;禁止写入 DS1302
              LCALL   WRITE
              CLR     00H
              CLR     01H
              CLR     ET1                 ;关闪中断
              CLR     TR1
WAITKEY000:   LCALL   DISPLAY             ;等待按键释放
              JNB     KEYSW0,WAITKEY000
              LCALL   DISPLAY
              JNB     KEYSW0,WAITKEY000
              LJMP    MAIN1
;
;**************** 闪动调时程序 ********************;
;
       INTT1: PUSH    ACC
              PUSH    PSW
              DJNZ    INTCON,GFLASHOUT
              MOV     INTCON,#CON_DATA
      GFLASH: CPL     00H
              JB      00H,GFLASH5
```

```
              MOV     72H,45H            ;全显示
              MOV     73H,44H
              MOV     75H,43H
              MOV     76H,42H
              MOV     78H,41H
              MOV     79H,40H
   GFLASHOUT: LCALL   DISPLAY
              POP     PSW
              POP     ACC
              RETI
;
   GFLASH5:   JB      01H,GFLASH6        ;调小时闪
              MOV     72H,#0AH
              MOV     73H,#0AH
              AJMP    GFLASHOUT
   GFLASH6:   MOV     75H,#0AH           ;调分钟闪
              MOV     76H,#0AH
              AJMP    GFLASHOUT
;
;****************加1程序***************;
;
   ADD1:      MOV     A,R7
              ADD     A,#01H
              DA      A
              RET
;
;****************分离程序********************;
;
   DIVIDE:    MOV     A,R0
              ANL     A,#0FH
              MOV     R1,A
              MOV     A,R0
              SWAP    A
              ANL     A,#0FH
              MOV     R2,A
              RET
;
;**************写 DS1302 程序****************;
;
   WRITE:     CLR     SCLK
              NOP
              SETB    RST
```

第 3 章 DS1302 实时时钟的设计

```
                NOP
                MOV     A,DS1302_ADDR
                MOV     R4,#8
        WRITE1: RRC     A                       ;送地址给 DS1302
                NOP
                NOP
                CLR     SCLK
                NOP
                NOP
                NOP
                MOV     IO,C
                NOP
                NOP
                NOP
                SETB    SCLK
                NOP
                NOP
                DJNZ    R4,WRITE1
                CLR     SCLK
                NOP
                MOV     A,DS1302_DATA
                MOV     R4,#8
        WRITE2: RRC     A
                NOP                             ;送数据给 DS1302
                CLR     SCLK
                NOP
                NOP
                MOV     IO,C
                NOP
                NOP
                NOP
                SETB    SCLK
                NOP
                NOP
                DJNZ    R4,WRITE2
                CLR     RST
                RET
;
;***************** 读 DS1302 程序 *******************;
;
        READ:   CLR     SCLK
                NOP
```

```
        NOP
        SETB    RST
        NOP
        MOV     A,DS1302_ADDR
        MOV     R4,#8
READ1:  RRC     A                       ;送地址给 DS1302
        NOP
        MOV     IO,C
        NOP
        NOP
        NOP
        SETB    SCLK
        NOP
        NOP
        NOP
        CLR     SCLK
        NOP
        NOP
        DJNZ    R4,READ1

        MOV     R4,#8
READ2:  CLR     SCLK
        NOP                             ;从 DS1302 中读出数据
        NOP
        NOP
        MOV     C,IO
        NOP
        NOP
        NOP
        NOP
        NOP
        RRC     A
        NOP
        NOP
        NOP
        NOP
        SETB    SCLK
        NOP
        DJNZ    R4,READ2
        MOV     DS1302_DATA,A
        CLR     RST
        RET
```

第 3 章 DS1302 实时时钟的设计

```
;
;************** 显示程序 *********************
        DISPLAY: MOV    R1,DISPFIRST
                 MOV    R5,#01H
        SPLAY:   MOV    A,R5
                 MOV    P2,A
                 MOV    A,@R1
                 MOV    DPTR,#TABS
                 MOVC   A,@A+DPTR
                 MOV    P0,A
                 MOV    A,R5
                 LCALL  DL1MS
                 INC    R1
                 MOV    A,R5
                 JB     ACC.7,ENDOUTS
                 RL     A
                 MOV    R5,A
                 AJMP   SPLAY
        ENDOUTS: MOV    P2,#00H
                 MOV    P0,#0FFH
                 RET
TABS: DB 0C0H,0F9H,0A4H,0B0H,99H,92H,82H,0F8H,80H,90H,0FFH,0C6H,0BFH,88H
;显示值 "0    1    2    3    4    5    6    7    8    9   不显   C    -    A"
;内存数 "0    1    2    3    4    5    6    7    8    9   0AH   0BH   0CH   0DH"
;************** 延时子程序 *********************
;1 ms 延时程序
        DL1MS: MOV    DELAYR6,#14H
        DL1:   MOV    DELAYR7,#19H
        DL2:   DJNZ   DELAYR7,DL2
               DJNZ   DELAYR6,DL1
               RET
;20 ms 延时程序
        DL20MS: CLR    BELL
                LCALL  DISPLAY
                LCALL  DISPLAY
                SETB   BELL
                RET
;延时程序
        DL05S:  MOV    DELAYR3,#20H
        DL05S1: LCALL  DISPLAY
                DJNZ   DELAYR3,DL05S1
                RET
```

END
;******************** 结束 **************************

3.6.2 C 源程序清单

```c
/*--------------------------------------
Real-Time Clock DS1302 program V9.1
MCU STC89C52RC   XAL 12 MHz
Build by Gavin Hu, 2010.6.16
--------------------------------------*/
#include <reg51.h>
//
#define uchar unsigned char
#define uint unsigned int
#define ulong unsigned long
sbit BUZZ = P3^7;
sbit KEY1 = P3^5;
sbit KEY2 = P3^6;
sbit CE = P1^3;
sbit SCLK = P1^1;
sbit IO = P1^2;
uchar hour_reg, minute_reg, second_reg;

/***********************************************************/
/* Prototypes */
/***********************************************************/
uchar    rbyte_3w();
void     reset_3w();
void     wbyte_3w(uchar);
void     read_time();
void     delay(uint);
void     display(uchar *);
void     time2str(uchar *);
void     time_set(void);

/*--------------------------------------
   main function
--------------------------------------*/
void main(void)
{
    uchar dispram[9];
    uchar s;
```

```
    reset_3w();
    wbyte_3w(0x8E);
    wbyte_3w(0x00);
    reset_3w();
    wbyte_3w(0x90);
    wbyte_3w(0xAB);
    reset_3w();
    wbyte_3w(0x81);
    s = rbyte_3w();
    reset_3w();
    if (s&0x80)
        {
        wbyte_3w(0x80);
        wbyte_3w(s&0x7f);
        reset_3w();
        }
    wbyte_3w(0x85);
    s = rbyte_3w();
    reset_3w();
    if (s&0x80)
        {
        wbyte_3w(0x84);
        wbyte_3w(s&0x7f);
        reset_3w();
        }
    while (1)
        {
        read_time();
        time2str(dispram);
        display(dispram);
        if (KEY1 == 0) time_set();
        }
}

/*---------------------------------------
    Time data to display string function
    Parameter:pointer of string
---------------------------------------*/
void time2str(uchar * ch)
{
ch[0] = hour_reg>>4;
ch[1] = hour_reg&0x0f;
ch[2] = 16;
```

```c
    ch[3] = minute_reg>>4;
    ch[4] = minute_reg&0x0f;
    ch[5] = 16;
    ch[6] = second_reg>>4;
    ch[7] = second_reg&0x0f;
}

/* ---------------------------------------
   Set time function
   ------------------------------------- */
void time_set(void)
{
uchar ch[8];
uchar i,c;
reset_3w();
wbyte_3w(0x80);
wbyte_3w(0x80);
reset_3w();
second_reg = 0;
time2str(ch);
do {
    display(ch);
    } while (KEY1 == 0);
c = 2;
while (c)
    {
    time2str(ch);
    if (c == 2) {ch[0]| = 0x80;ch[1]| = 0x80;}
        else {ch[3]| = 0x80;ch[4]| = 0x80;}
    display(ch);
    if (KEY1 == 0)
        {
        c--;
        do {
            display(ch);
            } while (KEY1 == 0);
        }
    if (KEY2 == 0)
        {
        if (c == 2)
            {
            hour_reg++;
            if ((hour_reg&0x0f)>9) hour_reg = (hour_reg&0xf0) + 0x10;
```

```
                    if (hour_reg>0x23) hour_reg = 0;
                    }
                    else
                    {
                    minute_reg++;
                    if ((minute_reg&0x0f)>9) minute_reg = (minute_reg&0xf0) + 0x10;
                    if (minute_reg>0x59) minute_reg = 0;
                    }
                for (i=0;i<50;i++) display(ch);
                }
        }
reset_3w();
wbyte_3w(0x84);
wbyte_3w(hour_reg);
reset_3w();
wbyte_3w(0x82);
wbyte_3w(minute_reg);
reset_3w();
wbyte_3w(0x80);
wbyte_3w(0x00);
reset_3w();
}

/*------------------------------------
  Delay function
  Parameter: unsigned int dt
  Delay time = dt(ms)
------------------------------------*/
void delay(unsigned int dt)
{
register unsigned char bt,ct;
for (; dt; dt--)
    for (ct=2;ct;ct--)
        for (bt=248; --bt; );
}

/*------------------------------------
  8 LED digital tubes display function
  Parameter: sting pointer to display
------------------------------------*/
void display(uchar * disp_ram)
{
static uchar disp_count;
```

```c
unsigned char i,j;
unsigned char code table[] =

{0xc0,0xf9,0xa4,0xb0,0x99,0x92,0x82,0xf8,0x80,0x90,0x88,

0x83,0xc6,0xa1,0x86,0x8e,0xbf,0xff};
disp_count = (disp_count + 1)&0x7f;
for (i = 0;i < 8;i++ )
    {
    j = disp_ram[i];
    if (j&0x80) P0 = (disp_count > 32)? table[j&0x7f]:0xff;
        else P0 = table[j];
    P2 = 0x01 << i;
    delay(1);
    P0 = 0xff;
    P2 = 0;
    }
}

/* ------------------------------------
   Read time function
   ------------------------------------ */
void read_time()
{
reset_3w();
wbyte_3w(0xBF);
second_reg = rbyte_3w()&0x7f;
minute_reg = rbyte_3w()&0x7f;
hour_reg = rbyte_3w()&0x3f;
reset_3w();
}

void reset_3w()     /* ---- reset and enable the 3 - wire interface ---- */
{
    CE = 0;
    SCLK = 0;
    CE = 0;
    SCLK = 0;
    CE = 1;
}

void wbyte_3w(uchar W_Byte)     /* ---- write one byte to the device ---- */
{
```

```
    uchar i;
        for(i = 0; i<8; ++i)
        {
            SCLK = 0;
            IO = W_Byte & 0x01;
            SCLK = 1;
            W_Byte>> = 1;
        }
}
uchar    rbyte_3w()     /* ---- read one byte from the device ---- */
{
uchar i;
uchar R_Byte;
    IO = 1;
    for(i = 0; i<8; i++)
    {
        SCLK = 0;
        R_Byte>> = 1;
        if (IO) R_Byte |= 0x80;
        SCLK = 1;
    }
    return R_Byte;
}
```

第 4 章 数字温度计的设计

4.1 系统功能

数字温度计测温范围为-55~125 ℃,精度误差在 0.5 ℃以内,用 4 位共阳 LED 数码管直读显示,要求高位为 0 时不显示,低于 0 ℃时前面显示"—"。

4.2 设计方案

传统的测温元件有热电偶和热电阻,但热电偶和热电阻测出的一般都是电压,然后再转换成对应的温度,需要比较多的外部硬件支持,而硬件电路复杂,软件调试也复杂,因此制作成本高。数字温度计设计可以采用美国 DALLAS 半导体公司继 DS1820 之后推出的一种改进型智能温度传感器 DS18B20 作为检测元件,测温范围为-55~125 ℃,分辨率最大可达 0.062 5 ℃。DS18B20 可以直接读出被测温度值(不用校准),而且采用单线与单片机通信,减少了外部的硬件电路,具有高精度和易使用的特点。

按照系统功能的要求,数字温度计由主控制器、测温单元、显示电路共 3 个模块组成。总体系统结构框图如图 4.1 所示。

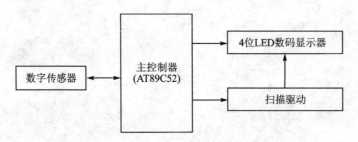

图 4.1 数字温度计系统结构框图

4.3 系统硬件仿真电路的设计

数字温度计的 Proteus 硬件仿真电路如图 4.2 所示,控制器使用 AT89C52 单片机,温度传感器使用 DS18B20,用 4 位共阳 LED 数码管以动态扫描法实现温度显示,从 P0 口输出段码,列扫描用 P2 口来实现,列驱动用 74HC244 来实现,可直接作为

LED 段码灯的电源。

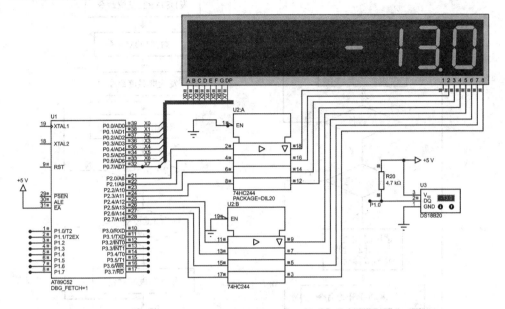

图 4.2　数字温度计的 Proteus 硬件仿真电路图

4.4　系统程序的设计

系统程序主要包括主程序、读出温度子程序、温度转换命令子程序、计算温度子程序、显示数据刷新子程序等。

1. 主程序

主程序的主要功能是负责温度的实时显示,读出并处理 DS18B20 的测量温度值。温度测量每 1 s 进行一次,其程序流程图如图 4.3 所示。

2. 读出温度子程序

读出温度子程序的主要功能是读出 DS18B20 RAM 中的 9 个字节,在读出时需进行 CRC 校验,校验有错时不进行温度数据的改写,其程序流程图如图 4.4 所示。

3. 温度转换命令子程序

温度转换命令子程序主要是发温度转换开始命令,当采用 12 位分辨率时,转换时间约为 750 ms,在本程序设计中采用 1 s 显示程序延时法等待转换的完成。温度转换命令子程序流程图如图 4.5 所示。

4. 计算温度子程序

计算温度子程序将 DS18B20 RAM 中的读取值进行 BCD 码的转换运算,并进行温度值正负的判定,其程序流程图如图 4.6 所示。

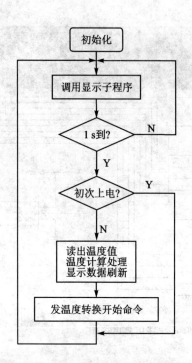
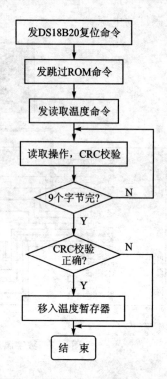

图 4.3　DS18B20 温度计主程序流程图　　　图 4.4　读出温度子程序流程图

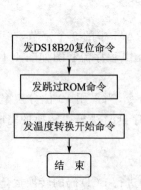
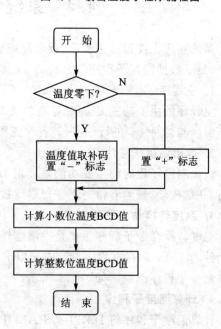

图 4.5　温度转换命令子程序流程图　　　图 4.6　计算温度子程序流程图

5. 显示数据刷新子程序

显示数据刷新子程序主要是对显示缓冲器中的显示数据进行刷新操作,当最高数据显示位为0时,将符号显示位移入下一位。程序流程图如图4.7所示。

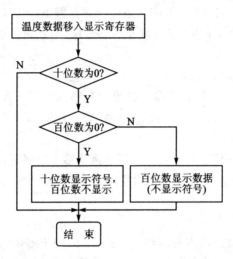

图 4.7　显示数据刷新子程序流程图

6. DS18B20 中的 ROM 命令

① Read ROM [33H]。

这个命令允许总线控制器读到 DS18B20 的 8 位系列编码、唯一的序列号和 8 位 CRC 码。只有在总线上存在单只 DS18B20 时才能使用这个命令。如果总线上有不止一个从机,那么当所有从机试图同时传送信号时,就会发生数据冲突(漏极开路连在一起形成相"与"的效果)。

② Match ROM [55H]。

这个是匹配 ROM 命令,后跟 64 位 ROM 序列,让总线控制器在多点总线上定位一只特定的 DS18B20。只有和 64 位 ROM 序列完全匹配的 DS18B20,才能响应随后的存储器操作;所有和 64 位 ROM 序列不匹配的从机,都将等待复位脉冲。这条命令在总线上有单个或多个器件时都可以使用。

③ Skip ROM [0CCH]。

这条命令允许总线控制器不用提供 64 位 ROM 编码就可以使用存储器操作命令,在单线总线情况下,可以节省时间。如果总线上有不止一个从机,在 Skip ROM 命令之后跟着发一条读命令,那么由于多个从机同时传送信号,总线上就会发生数据冲突(漏极开路下拉效果相当于相"与")。

④ Search ROM [0F0H]。

当一个系统初次启动时,总线控制器可能并不知道单线总线上有多少器件或它们的 64 位 ROM 编码,而搜索 ROM 命令允许总线控制器用排除法识别总线上的所有从机的 64 位编码。

⑤ Alarm Search [0ECH]。

这条命令的流程和 Search ROM 相同,但是,只有在最近一次测温后遇到符合报警条件的情况下,DS18B20 才会响应这条命令。报警条件定义为温度高于 TH 或低于 TL。只要 DS18B20 不掉电,报警状态将一直保持,直到再一次测得的温度值达不到报警条件。

⑥ Write Scratchpad [4EH]。

这个命令向 DS18B20 的暂存器 TH 和 TL 写入数据。控制器可以在任意时刻发出复位命令来中止写入。

⑦ Read Scratchpad [0BEH]。

这个命令用于读取暂存器的内容。读取将从第 1 个字节开始,一直进行下去,直到第 9(CRC)个字节读完。如果不想读完所有字节,控制器可以在任意时间发出复位命令来中止读取。

⑧ Copy Scratchpad [48H]。

这个命令把暂存器的内容复制到 DS18B20 的 EEPROM 存储器中,即把温度报警触发字节存入非易失性存储器中。如果总线控制器在这条命令之后跟着发出读数据命令,而 DS18B20 又忙于把暂存器的内容复制到 EEPROM 存储器中,那么 DS18B20 就会输出一个"0";如果复制结束,则 DS18B20 输出"1"。如果使用寄生电源,则总线控制器必须在这条命令发出后立即启动强上拉,并至少保持 10 ms。

⑨ Convert T [44H]。

这条命令启动一次温度转换而无须其他数据。温度转换命令被执行后,DS18B20 保持等待状态。如果总线控制器在这条命令之后跟着发出读数据命令,而 DS18B20 又忙于做时间转换,那么 DS18B20 将在总线上输出"0";若温度转换完成,则输出"1"。如果使用寄生电源,则总线控制器必须在发出这条命令后立即启动强上拉,并保持 500 ms 以上的时间。

⑩ Recall E^2 [0B8H]。

这条命令把报警触发器中的值复制回暂存器。这种复制操作在 DS18B20 上电时将自动执行,这样器件一上电,暂存器中马上就存在有效的数据了。若在这条命令发出之后发出读数据命令,那么器件会输出温度转换忙的标识:"0"=忙,"1"=完成。

⑪ Read Power Supply [0B4H]。

若把这条命令发给 DS18B20 后发出读时间隙,器件会返回它的电源模式:"0"=寄生电源,"1"=外部电源。

7. 温度数据的计算处理方法

从 DS18B20 读取出的二进制值必须先转换成十进制 BCD 码,才能用于字符的显示。因为 DS18B20 的转换精度为 9～12 位可选的,所以为了提高精度可采用 12 位转换精度。在采用 12 位转换精度时,温度寄存器中的值是以 0.062 5 为步进的，即温度值为温度寄存器中的二进制值乘以 0.062 5，就是实际的十进制温度值。表 4.1 所列是 DS18B20 温度与二进制及十六进制表示值的对应关系，从表中可知，一个十进制温度值和二进制值之间有很明显的关系，就是把二进制高字节的低半字节和低字节的高半字节组成一个字节，将这个字节的二进制值转化为十进制 BCD 码后，就是温度值的百、十、个位值，而剩下的低字节的低半字节转化成十进制后，就是温度值的小数部分。因为小数部分是半个字节，十六进制值的范围是 0～F，所以转换成十进制小数值就是 0.062 5 的倍数（0～15 倍）。此时，需用 4 位的数码管来显示小数部分，而在实际应用中不必有这么高的精度，设计中一般采用 1 位数码管来显示小数，可以精确到 0.1 ℃。表 4.2 所列是小数部分十六进制和十进制的近似对应关系。

表 4.1　DS18B20 温度与表示值对应表

温度/℃	二进制表示	十六进制表示
125	0000 0111　1101 0000	07D0H
85	0000 0101　0101 0000	0550H
25.062 5	0000 0001　1001 0001	0191H
10.125	0000 0000　1010 0010	00A2H
0.5	0000 0000　0000 1000	0008H
0	0000 0000　0000 0000	0000H
−0.5	1111 1111　1111 1000	FFF8H
−10.125	1111 1111　0101 1110	FF5EH
−25.062 5	1111 1110　0110 1111	FE6FH
−55	1111 1100　1001 0000	FC90H

表 4.2　小数部分十六进制和十进制的近似对应关系表

小数部分十六进制值	0	1	2	3	4	5	6	7	8	9	A	B	C	D	E	F
十进制小数近似值	0	0	1	1	2	3	3	4	5	5	6	6	7	8	8	9

4.5　软件调试与运行结果

系统的调试以程序为主，可先编写一个测试小程序来判断仿真硬件电路是否正常，然后分别进行显示程序、主程序、读出温度子程序、温度转换命令子程序、计算温度子程序、显示数据刷新等子程序的编程及调试。由于 DS18B20 与单片机采用串行数据传送，因此，对 DS18B20 进行读/写编程时必须严格地保证读/写时序，否则将无法读取测量结果。

4.6 源程序清单

4.6.1 汇编源程序清单

```
;****************************************************
;                课程设计程序:数字温度计              *
;            显示精度 0.1 ℃,测温范围 -55～125 ℃      *
;                  AT89C52,12 MHz 晶振                *
;****************************************************
;
;****************************************************
;           常数定义
;
;****************************************************
    TIMEL       EQU     0E0H    ;定时器 T0 的 20 ms 时间常数
    TIMEH       EQU     0B1H    ;定时器 T0 的 20 ms 时间常数
    TEMPHEAD    EQU     36H     ;DS18B20 读出字节存放首址(共读 9 个字节)
;
;****************************************************
;           工作内存定义
;
;****************************************************
    BITST       DATA    20H     ;用作标志位
    TIME1SOK    BIT     BITST.1 ;1 s 定时时间标志,1 s 到时为 1
    TEMPONEOK   BIT     BITST.2 ;上电标志,刚上电为 0,读出一次后为 1
    TEMPL       DATA    26H     ;读出温度低字节存放——整数低 4 位 + 小数位 4 位
    TEMPH       DATA    27H     ;读出温度高字节存放——4 位符号位 + 整数高 4 位
    TEMPHC      DATA    28H     ;用于存放处理好的 BCD 码温度值:百位 + 十位
    TEMPLC      DATA    29H     ;用于存放处理好的 BCD 码温度值:个位 + 小数位
;
;
;****************************************************
;           引脚定义
;****************************************************
    TEMPDIN     BIT     P1.0    ;DS18B20 数据接口
;****************************************************
;           中断向量区
;****************************************************
```

```
            ORG     0000H
            LJMP    START
            ORG     000BH
            LJMP    T0IT
;************************************************
;       系统初始化
;************************************************
            ORG     100H
START:      MOV     SP,#60H             ;堆栈底
CLSMEM:     MOV     R0,#20H
            MOV     R1,#60H             ;20H~7FH 清零
CLSMEM1:    MOV     @R0,#00H
            INC     R0
            DJNZ    R1,CLSMEM1
;
            MOV     TMOD,#00100001B     ;定时器 T0 作方式 1（16 位）
            MOV     TH0,#TIMEL          ;装 20 ms 定时初值
            MOV     TL0,#TIMEH          ;装 20 ms 定时初值
            MOV     P2,#00H             ;LED 显示关
            SJMP    INIT
;
ERROR:      NOP
            LJMP    START
;
            NOP
INIT:       NOP
            SETB    ET0
            SETB    TR0
            SETB    EA
            MOV     PSW,#00H
            CLR     TEMPONEOK           ;第一次上电为 0
            LJMP    MAIN
;
;************************************************
;       定时器 T0 中断服务程序
;************************************************
T0IT:       PUSH    PSW
            MOV     PSW,#10H
            MOV     TH0,#TIMEH
            MOV     TL0,#TIMEL
            INC     R7
            CJNE    R7,#32H,T0IT1
```

```
                MOV     R7, #00H
                SETB    TIME1SOK            ;1 s 定时到标志位为 1
TOIT1:          POP     PSW
                RETI
;
;****************************************************
;       主程序
;****************************************************

MAIN:           LCALL   DISP1               ;调用显示子程序
                JNB     TIME1SOK, MAIN
                CLR     TIME1SOK            ;测温每 1 s 一次
                JNB     TEMPONEOK, MAIN2    ;上电时先温度转换一次
                LCALL   READTEMP1           ;读出温度值子程序
                LCALL   CONVTEMP            ;温度 BCD 码计算处理子程序
                LCALL   DISPBCD             ;显示区 BCD 码温度值刷新子程序
                LCALL   DISP1               ;消闪烁,显示一次
MAIN2:          LCALL   READTEMP            ;温度转换开始
                SETB    TEMPONEOK
                LJMP    MAIN
;
;****************************************************
;       以下为子程序区
;****************************************************
;       DS18B20 复位子程序
;****************************************************
INITDS1820:     SETB    TEMPDIN
                NOP
                NOP
                CLR     TEMPDIN
                MOV     R6, #0A0H           ;DELAY 480 μs
                DJNZ    R6, $
                MOV     R6, #0A0H
                DJNZ    R6, $
                SETB    TEMPDIN
                MOV     R6, #32H            ;DELAY 70 μs
                DJNZ    R6, $
                MOV     R6, #3CH
LOOP1820:       MOV     C, TEMPDIN
                JC      INITDS1820OUT
                DJNZ    R6, LOOP1820
                MOV     R6, #064H           ;DELAY 200 μs
```

```
                    DJNZ    R6, $
                    SJMP    INITDS1820
                    RET
;
INITDS1820OUT:      SETB    TEMPDIN
                    RET
;
;*************************************************************
;       读 DS18B20 的程序，从 DS18B20 中读出一个字节的数据
;*************************************************************
READDS1820:         MOV     R7, #08H
                    SETB    TEMPDIN
                    NOP
                    NOP
READDS1820LOOP:     CLR     TEMPDIN
                    NOP
                    NOP
                    NOP
                    SETB    TEMPDIN
                    MOV     R6, #07H            ;DELAY 15 μs
                    DJNZ    R6, $
                    MOV     C, TEMPDIN
                    MOV     R6, #3CH            ;DELAY 120 μs
                    DJNZ    R6, $
                    RRC     A
                    SETB    TEMPDIN
                    DJNZ    R7, READDS1820LOOP
                    MOV     R6, #3CH            ;DELAY 120 μs
                    DJNZ    R6, $
                    RET
;
;
;*************************************************************
;       写 DS18B20 的程序，从 DS18B20 中写一个字节的数据
;*************************************************************
WRITEDS1820:        MOV     R7, #08H
                    SETB    TEMPDIN
                    NOP
                    NOP
WRITEDS1820LOP:     CLR     TEMPDIN
                    MOV     R6, #07H            ;DELAY 15 μs
                    DJNZ    R6, $
```

```
                RRC     A
                MOV     TEMPDIN, C
                MOV     R6, #34H          ;DELAY 104 μs
                DJNZ    R6, $
                SETB    TEMPDIN
                DJNZ    R7, WRITEDS1820LOP
                RET
;
;****************************************************
;      以下为读温度子程序
;****************************************************
READTEMP:       LCALL   INITDS1820        ;温度转换开始命令程序
                MOV     A, #0CCH
                LCALL   WRITEDS1820       ;SKIP ROM
                MOV     R6, #34H          ;DELAY 104 μs
                DJNZ    R6, $
                MOV     A, #44H
                LCALL   WRITEDS1820       ;START CONVERSION
                MOV     R6, #34H          ;DELAY 104 μs
                DJNZ    R6, $
                RET
;
READTEMP1:      LCALL   INITDS1820        ;读出温度字节子程序
                MOV     A, #0CCH
                LCALL   WRITEDS1820       ;SKIP ROM
                MOV     R6, #34H          ;DELAY 104 μs
                DJNZ    R6, $
                MOV     A, #0BEH
                LCALL   WRITEDS1820       ;SCRATCHPAD
                MOV     R6, #34H          ;DELAY 104 μs
                DJNZ    R6, $
                MOV     R5, #09H          ;读 9 个字节
                MOV     R0, #TEMPHEAD
                MOV     B, #00H
READTEMP2:      LCALL   READDS1820
                MOV     @R0, A
                INC     R0
READTEMP21:     LCALL   CRC8CAL           ;CRC 校验
                DJNZ    R5, READTEMP2
                MOV     A, B
                JNZ     READTEMPOUT       ;校验出错结束
                MOV     A, TEMPHEAD + 0   ;校验正确将前 2 个字节(温度)暂存
```

第4章 数字温度计的设计

```
                MOV     TEMPL, A
                MOV     A, TEMPHEAD + 1
                MOV     TEMPH, A
READTEMPOUT:    RET
;
;
;*******************************************************
;       处理温度 BCD 码子程序
;*******************************************************
CONVTEMP:       MOV     A, TEMPH
                ANL     A, #80H
                JZ      TEMPC1
                CLR     C                       ;负温度的补码处理(+1后取反)
                MOV     A, TEMPL
                CPL     A
                ADD     A, #01H
                MOV     TEMPL, A
                MOV     A, TEMPH
                CPL     A
                ADDC    A, #00H
                MOV     TEMPH, A                ;TEMPHC HI = 符号位
                MOV     TEMPHC, #0BH
                SJMP    TEMPC11
;
TEMPC1:         MOV     TEMPHC, #0AH            ;正温度处理
TEMPC11:        MOV     A, TEMPHC
                SWAP    A
                MOV     TEMPHC, A
                MOV     A, TEMPL
                ANL     A, #0FH                 ;温度小数位处理(乘以0.0625)
                MOV     DPTR, #TEMPDOTTAB       ;用查表处理小数位
                MOVC    A, @A + DPTR
                MOV     TEMPLC, A               ;TEMPLC LOW = 小数部分 BCD
;
                MOV     A, TEMPL                ;处理温度整数部分
                ANL     A, #0F0H
                SWAP    A
                MOV     TEMPL, A
                MOV     A, TEMPH
                ANL     A, #0FH
                SWAP    A
                ORL     A, TEMPL
```

```
            LCALL   HEX2BCD1
            MOV     TEMPL, A
            ANL     A, #0F0H
            SWAP    A
            ORL     A, TEMPHC           ;TEMPHC LOW 放十位数 BCD
            MOV     TEMPHC, A
            MOV     A, TEMPL
            ANL     A, #0FH
            SWAP    A                   ;TEMPLC HI 放个位数 BCD
            ORL     A, TEMPLC
            MOV     TEMPLC, A
            MOV     A, R7
            JZ      TEMPC12
            ANL     A, #0FH
            SWAP    A
            MOV     R7, A
            MOV     A, TEMPHC           ;TEMPHC HI 放百位数 BCD
            ANL     A, #0FH
            ORL     A, R7
            MOV     TEMPHC, A
TEMPC12:    RET
;
;*********************************************************
;    小数部分码表
;*********************************************************
TEMPDOTTAB: DB      00H, 01H, 01H, 02H, 03H, 03H, 04H, 04H, 05H, 06H
;
            DB      06H, 07H, 08H, 08H, 09H, 09H
;
            RET
;
;*********************************************************
;    显示区 BCD 码温度值刷新子程序
;*********************************************************
; 低于 0 ℃要显示"-",高位 0 不显示
DISPBCD:    MOV     A, TEMPLC
            ANL     A, #0FH
            MOV     70H, A
            MOV     A, TEMPLC
            SWAP    A
            ANL     A, #0FH
            MOV     71H, A
```

第4章 数字温度计的设计

```
                MOV     A, TEMPHC
                ANL     A, #0FH
                MOV     72H, A
                MOV     A, TEMPHC
                SWAP    A
                ANL     A, #0FH
                MOV     73H, A
                MOV     A, TEMPHC
                ANL     A, #0F0H
                CJNE    A, #010H, DISPBCD0
                SJMP    DISPBCD2
;
DISPBCD0:       MOV     A, TEMPHC
                ANL     A, #0FH
                JNZ     DISPBCD2            ;十位数是0
                MOV     A, TEMPHC
                SWAP    A
                ANL     A, #0FH
                MOV     73H, #0AH           ;符号位不显示
                MOV     72H, A              ;十位数显示符号
DISPBCD2:       RET
;
;*****************************************************
;                   显示子程序
;*****************************************************
;显示数据在70H~73H单元内,用4位共阳数码管,P0口输出段码数据
;P2口作扫描控制,每个LED数码管亮1 ms再逐位循环
;
DISP1:          MOV     R1, #70H            ;指向显示数据首址
                MOV     R5, #80H            ;扫描控制字初值
PLAY:           MOV     P0, #0FFH
                MOV     A, R5               ;扫描字放入A
                MOV     P2, A               ;从P2口输出
                MOV     A, @R1              ;取显示数据到A
                MOV     DPTR, #TAB          ;取段码表地址
                MOVC    A, @A+DPTR          ;查显示数据对应段码
                MOV     P0, A               ;段码放入P0口
                MOV     A, R5
                JNB     ACC.6, LOOP5        ;小数点处理
                CLR     P0.7
LOOP5:          LCALL   DL1MS               ;显示1 ms
                INC     R1                  ;指向下一地址
```

```
                MOV     A,R5                ;扫描控制字放入A
                JB      ACC.4,ENDOUT        ;ACC.5=0时一次显示结束
                RR      A                   ;A中数据循环左移
                MOV     R5,A                ;放回R5内
                AJMP    PLAY                ;跳回PLAY循环
ENDOUT:         MOV     P0,#0FFH            ;一次显示结束,P0口复位
                MOV     P2,#00H             ;P2口复位
                RET                         ;子程序返回
TAB:    DB      0C0H,0F9H,0A4H,0B0H,99H,92H,82H,0F8H,80H,90H,0FFH,0BFH
;共阳段码表      "0"   "1"   "2"   "3"  "4"  "5"  "6"  "7"  "8"  "9"  "不亮" "-"
;
;1 ms延时程序(LED显示程序用)
DL1MS:          MOV     R6,#14H
DL1:            MOV     R7,#19H
DL2:            DJNZ    R7,DL2
                DJNZ    R6,DL1
                RET
;
;****************************************************
;       单字节十六进制转BCD
;****************************************************
HEX2BCD1:       MOV     B,#064H             ;十六进制→BCD
                DIV     AB                  ;B=A%100
                MOV     R7,A                ;R7=百位数
                MOV     A,#0AH
                XCH     A,B
                DIV     AB                  ;B=A%B
                SWAP    A
                ORL     A,B
                RET
;
;
;****************************************************
;       CRC校验程序
;       X^8 + X^5 + X^4 + 1
;****************************************************
CRC8CAL:        PUSH    ACC
                MOV     R7,#08H
;
CRC8LOOP1:      XRL     A,B
                RRC     A
```

```
                MOV     A, B
                JNC     CRC8LOOP2
                XRL     A, #18H
;
CRC8LOOP2:      RRC     A
                MOV     B, A
                POP     ACC
                RR      A
                PUSH    ACC
                DJNZ    R7, CRC8LOOP1
                POP     ACC
                RET
;
                END                             ;程序结束
```

4.6.2　C 源程序清单

```c
/* --------------------------------------
DS18B20 Digital Thermometer program V10.1
MCU STC89C52RC   XAL 12 MHz
Build by Gavin Hu, 2010.6.14
-------------------------------------- */
#include "reg51.h"
#include "intrins.h"
#define NO_DISPLAY 17
#define DISP_SIGN 16
sbit ONE_WIRE_DQ = P1^0;

void delay_ms(unsigned int);
void delay_us(register unsigned char);
void temp2str(signed int tmep,unsigned char *);
void display(unsigned char *);
void start_convert(void);
signed int read_temperature(void);
unsigned char OW_reset(void);
unsigned char OW_read_byte(void);
void OW_write_byte(unsigned char val);
void main()
{
unsigned char i;
unsigned char dispram[8];
for (i = 0;i < 8;i++) dispram[i] = NO_DISPLAY;
```

```c
    while (1)
       {
          start_convert();
          for (i = 0; i < 120; i ++ ) display(dispram);
          temp2str(read_temperature(),dispram);
       }
}
/* ----------------------------------------
// Start DS18B20 Temperature Convert
------------------------------------- */
void start_convert(void)
{
OW_reset();
OW_write_byte(0xCC);  //Skip ROM
OW_write_byte(0x44);  // Start Conversion
}
/* ----------------------------------------
// Read Temperature
// returns the Temperature
------------------------------------- */
signed int read_temperature(void)
{
unsigned char get[9];
signed int temp;
unsigned char i;
OW_reset();
OW_write_byte(0xCC);              //Skip ROM
OW_write_byte(0xBE);              //Read Scratch Pad
for (i = 0; i < 9; i ++ ) get[i] = OW_read_byte();
temp = get[1];                    //Sign byte + lsbit
temp = (temp<<8) | get[0];        //Temp data plus lsb
return temp;
}
/* ----------------------------------------
// OW_RESET - performs a reset on the one-wire bus and
// returns the presence detect
------------------------------------- */
unsigned char OW_reset(void)
{
unsigned char presence;
ONE_WIRE_DQ = 0;                  //pull ONE_WIRE_DQ line low
delay_us(240);                    //leave it low for 480 μs
```

第 4 章 数字温度计的设计

```c
    ONE_WIRE_DQ = 1;              //allow line to return high
    delay_us(33);                 //wait for presence 70 μs
    presence = !ONE_WIRE_DQ;      //get presence signal
    delay_us(205);                //wait for end of timeslot
    return (presence);            //presence signal returned
    }                             //0 = presence, 1 = no part

/* ----------------------------------------
// READ_BYTE - reads a byte from the one-wire bus
----------------------------------------*/
unsigned char OW_read_byte(void)
{
unsigned char i;
unsigned char value;
for (i = 0;i < 8;i ++)
    {
    ONE_WIRE_DQ = 0;              //pull ONE_WIRE_DQ low to start timeslot
    value >>= 1;                  //delay
    ONE_WIRE_DQ = 1;              //then return high
    delay_us(1);                  //delay 6μs from start of timeslot
    if (ONE_WIRE_DQ) value |= 0x80; //reads byte in, one byte at a time and then
    delay_us(25);                 //delay 55 μs
    }
//delay_us(60);
return(value);
}
/* ----------------------------------------
// WRITE_BYTE - writes a byte to the one-wire bus
----------------------------------------*/
void OW_write_byte(char val)
{
unsigned char i;
for (i = 0; i < 8; i ++)          //writes byte, one bit at a time
    {
    ONE_WIRE_DQ = 0;              //pull ONE_WIRE_DQ low to start timeslot
    delay_us(1);                  //delay 6 μs
    ONE_WIRE_DQ = val&0x01;       //return ONE_WIRE_DQ high if write 1
    delay_us(30);                 //hold value for remainder of timeslot 64us
    ONE_WIRE_DQ = 1;
    val >>= 1;                    //shifts val right "i"
```

spaces
        }
}
/* --
   Temperature data to display string function
   Parameter: int temp, pointer of string
   -- */
```c
void temp2str(signed int temp, unsigned char * ch)
{
unsigned char sign;
if (temp<0)
    {
    sign = 1;
    temp = (~temp) + 1;
    }
    else sign = 0;
ch[7] = ((temp&0x000f) * 10 + 8)/16;
temp>> = 4;
ch[6] = (temp % 10)|0x40;
temp / = 10;
ch[5] = temp % 10;
temp / = 10;
ch[4] = temp % 10;
ch[3] = NO_DISPLAY;
if (ch[4] == 0)
    {
    ch[4] = NO_DISPLAY;
    if (ch[5] == 0) ch[5] = NO_DISPLAY;
    }
if (sign)
    {
    if (ch[5] == NO_DISPLAY) ch[5] = DISP_SIGN;
        else if (ch[4] == NO_DISPLAY) ch[4] = DISP_SIGN;
        else ch[3] = DISP_SIGN;
    }
}
```
/* --
   8 LED digital tubes display function
   Parameter: sting pointer to display
   -- */
```c
void display(unsigned char * disp_ram)
{
```

```c
    static unsigned char disp_count;
    unsigned char i;
    unsigned char code table[] =

    {0xc0,0xf9,0xa4,0xb0,0x99,0x92,0x82,0xf8,0x80,0x90,0x88,

    0x83,0xc6,0xa1,0x86,0x8e,0xbf,0xff};
    disp_count = (disp_count + 1)&0x7f;
    for (i = 0;i<8;i++)
        {
        if (disp_ram[i]&0x80) P0 = (disp_count>32)? table[disp_ram[i]&0x3f]:0xff;
                else P0 = table[disp_ram[i]&0x3f];
        if (disp_ram[i]&0x40) P0 &= 0x7f;
        P2 = 0x01<<i;
        delay_ms(1);
        P0 = 0xff;
        P2 = 0;
        }
}

/* ---------------------------------
    Delay function
    Parameter: unsigned char dt
    Delay time = dt * 2 + 5(μs)
    ----------------------------------*/
void delay_us(register unsigned char dt)
{
while (--dt);
}


/* ---------------------------------
    Delay function
    Parameter: unsigned int dt
    Delay time = dt(ms)
    ----------------------------------*/
void delay_ms(unsigned int dt)
{
register unsigned char bt,ct;
for (;dt;dt--)
    for (ct = 2;ct;ct--)
        for (bt = 250; --bt; );
}
```

第 5 章 低频信号发生器的设计

5.1 系统功能

低频信号发生器要求能输出 0.1～50 Hz 的正弦波、三角波信号,其中,正弦波和三角波信号可以用按键选择输出,输出信号的频率可以在 0.1～50 Hz 的范围内调整。

5.2 设计方案

因为输出信号的频率较低,所以可以使用单片机作为信号数据产生源,用中断查表法完成波形数据的输出,再用 D/A 转换器输出规定的波形信号。另外,也可以利用多余的端口经 D/A 转换器输出 0°～360°的移相波形,同时也可以输出一路方波信号。系统实现的结构框图如图 5.1 所示。

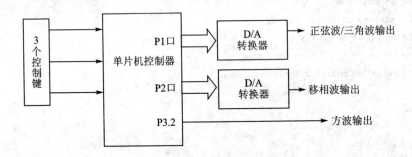

图 5.1 低频信号源的系统结构框图

5.3 系统硬件仿真电路的设计

低频信号源的 Proteus 硬件仿真电路如图 5.2 所示。

1. 控制部分

控制芯片选择 AT89C52 单片机。P3.3～P3.5 口接 3 个按键,其中,P3.3 口按键为频率增加键,P3.4 口按键为频率减小相键,P3.5 口按键为正弦波与三角波选择按键。P1 口输出正弦波或三角波数据,P2 口输出移相波数据,P3.2 输出方波。

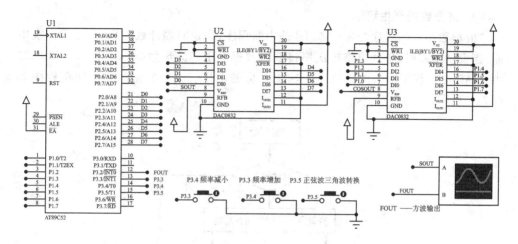

图 5.2　低频信号源的 Proteus 硬件仿真电路图

2. 数/模(D/A)转换部分

DAC0832 是 CMOS 工艺制造的 8 位 D/A 转换器,属于 8 位电流输出型 D/A 转换器,转换时间为 1 μs,片内带输入数字锁存器。DAC0832 与单片机接成数据直接写入方式,当单片机把一个数据写入 DAC 寄存器时,DAC0832 的输出模拟电压信号将随之对应变化。利用 D/A 转换器可以产生各种波形,如方波、三角波、锯齿波等,以及由它们组合产生的复合波形和不规则波形,这些复合波形利用标准的测试设备是很难产生的。

5.4　系统程序的设计

1. 初始化程序

初始化程序的主要工作是设置定时器的工作模式、初值预置、开中断、打开定时器等。在这里,定时器 T0 工作于 16 位定时模式,单片机按定时时间重复地把波形数据送到 DAC0832 的寄存器。初始化程序流程图如图 5.3 所示。

2. 按键扫描程序

按键扫描程序的任务是检查 3 个按键是否按下,如果按下,则执行相应的功能。这里,3 个按键分别用于频率增加、频率减小和正弦波与三角波的选择。其程序流程图如图 5.4 所示。

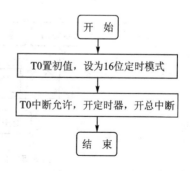

图 5.3　初始化程序流程图

3. 波形数据产生程序

波形数据产生程序是定时器 T0 的中断程序,当定时器计数溢出时,发生一次中断;当发生中断时,单片机将按次序将波形数据表中的波形数据一一送入 DAC0832 中,DAC0832 根据输入的数据大小输出对应的电压。波形数据产生程序流程图如图 5.5 所示。

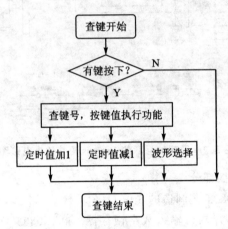

图 5.4 按键扫描程序流程图

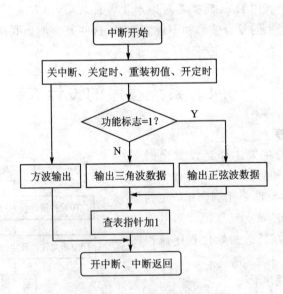

图 5.5 波形数据产生程序流程图

4. 主程序

主程序的任务是进行上电初始化,并在程序运行中不断查询按键情况执行相应的功能。

5.5 软件调试与运行结果

Proteus 硬件仿真电路的调试较简单,只要元器件连线无误,一般均能一次成功。软件的调试主要是各子程序的调试,其中,频率的增减按键因计数器为16位定时器,最大值为65 535,加或减一变化很慢,所以可在加减时用255作为加减数,这样频率的调整变化会快一些,但是在接近最高频率时会变化太快。如果加减时用1作为加减数,那么在频率的高端变化平稳,而在频率的低端则变化太慢。在调试时可根据应用特点选择加减数的大小。

简易低频信号源输出的频率不是很大,在设计中一周期波形由256个采样点合成,波形不是很光滑,如果增加采样点,则输出的频率会更低,所以在设计中应根据应用特点选择合理的采样点数。用单片机产生低频率信号的最大优点是,可以输出产生复杂的不规则波形,这是一般的通用信号源无法做到的。图5.6和图5.7所示分别为正弦波仿真及三角波仿真输出时的窗口图。

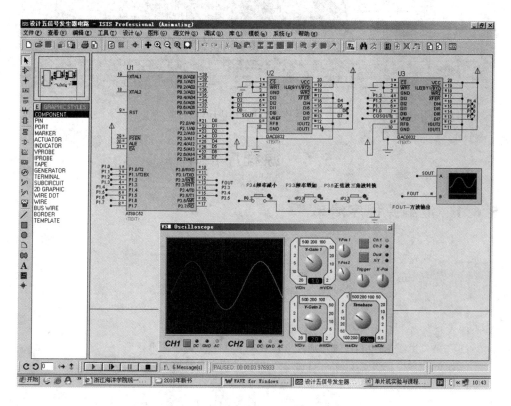

图 5.6 正弦波仿真输出时的窗口图

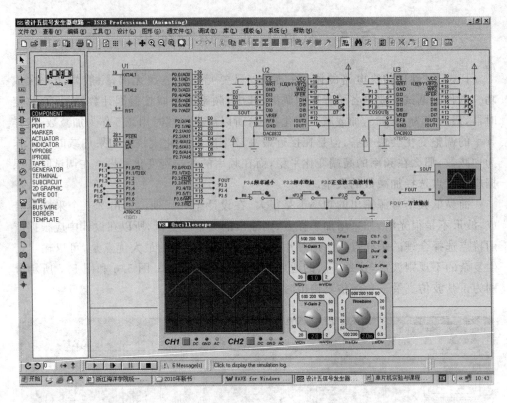

图 5.7 三角波仿真输出时的窗口图

5.6 源程序清单

5.6.1 汇编源程序清单

```
;******************************
;       课程设计程序:低频信号发生器      ;*
;           正弦波/三角波发生器          ;*
;              12 MHz 晶振              ;*
;******************************
;
;正弦波发生器,key0 口按键增加输出频率,key1 口按键减小输出频率
;sinout 口输出正弦波,cosout 口输出余弦波,使用定时器 T0,16 位定时模式
;R6、R7 用作 10 ms 延时寄存器,Fout 输出方波
;
key0        bit     P3.3            ;频率减小按键
key1        bit     P3.4            ;频率增加按键
key2        bit     P3.5            ;正弦波/三角波转换按键
```

```
        sinout    EQU    P2              ;正弦波/三角波输出
        ;cosout   EQU    P1              ;余弦波输出
        FOUT      BIT    P3.2            ;方波输出
        FLAG      BIT    00H             ;正弦波/三角波输出标志
        SINP      DATA   30H             ;正弦波查表指针
        COSP      DATA   31H             ;余弦波查表指针
        THOD      DATA   32H             ;定时器初值存放(高8位)
        TLOD      DATA   33H             ;定时器初值存放(低8位)
;
        ORG       0000H
        LJMP      START
        ORG       000BH
        LJMP      INTT0
;
        ORG       0060H
;
START:  MOV       SP,#70H
        MOV       SINP,#00H
        MOV       COSP,#40H
        MOV       TMOD,#11H
        MOV       THOD,#0FFH              ;初值,决定波形频率
        MOV       TLOD,#0AFH
        MOV       TH0,THOD
        MOV       TL0,TLOD
        clr       FLAG
CONZS:  CPL       FLAG                    ;0 正弦波,1 三角波
        JB        FLAG,NOSIN
        MOV       DPTR,#LIST
MAIN0:  SETB      ET0
        SETB      EA
        SETB      TR0
MAIN:   JNB       key0,INCKEY
        JNB       key1,DECKEY
        JNB       key2,CONSIN
;       ORL       PCON,#01H
        LJMP      MAIN
;
NOSIN:  MOV       DPTR,#LIST1
        AJMP      MAIN0
;按键功能,输出频率增大
INCKEY: LCALL     DL10MS
        JB        key0,MAIN
```

```
                MOV     A,TL0D
                CJNE    A,#0AFH,INC1
                LJMP    MAIN
INC1:           INC     TL0D
                LJMP    MAIN
;按键功能,输出频率减小
DECKEY:         LCALL   DL10MS
                JB      key1,MAIN
                MOV     A,TL0D
                CJNE    A,#00H,DEC1
                LJMP    MAIN
DEC1:           DEC     TL0D
                LJMP    MAIN
;按键功能,输出波形转换
CONSIN:         LCALL   DL10MS
                JB      key2,MAIN
WAITOFF:        JNB     key2,WAITOFF        ;等待按键释放
                LCALL   DL10MS
                JNB     key2,WAITOFF
                AJMP    CONZS

;定时器 T0 中断程序
INTT0:          PUSH    ACC
                CPL     Fout                ;方波输出,作辅助功能用
                MOV     TH0,TH0D
                MOV     TL0,TL0D
                MOV     A,SINP
                MOVC    A,@A+DPTR
                MOV     sinout,A            ;正弦波从 sinout 口输出
;               MOV     A,COSP
;               MOVC    A,@A+DPTR
;               MOV     cosout,A            ;余弦波从 cosout 口输出
                INC     SINP
;               INC     COSP
                POP     ACC
                RETI
;10 ms 延时程序
DL512:          MOV     R7,#0FFH
LOOP:           DJNZ    R7,LOOP
                RET
DL10MS:         MOV     R6,#10
LOOP1:          LCALL   DL512
```

```
        DJNZ      R6,LOOP1
                  RET
```
;正弦函数表(共256个点,每点1.406 25°)
```
LIST:   DB   80H,83H,85H,88H,8AH,8DH,8FH,92H
        DB   94H,97H,99H,9BH,9EH,0A0H,0A3H,0A5H
        DB   0A7H,0AAH,0ACH,0AEH,0B1H,0B3H,0B5H,0B7H
        DB   0B9H,0BBH,0BDH,0BFH,0C1H,0C3H,0C5H,0C7H
        DB   0C9H,0CBH,0CCH,0CEH,0D0H,0D1H,0D3H,0D4H
        DB   0D6H,0D7H,0D8H,0DAH,0DBH,0DCH,0DDH,0DEH
        DB   0DFH,0E0H,0E1H,0E2H,0E3H,0E3H,0E4H,0E4H
        DB   0E5H,0E5H,0E6H,0E6H,0E7H,0E7H,0E7H,0E7H
        DB   0E7H,0E7H,0E7H,0E7H,0E6H,0E6H,0E5H,0E5H
        DB   0E4H,0E4H,0E3H,0E3H,0E2H,0E1H,0E0H,0DFH
        DB   0DEH,0DDH,0DCH,0DBH,0DAH,0D8H,0D7H,0D6H
        DB   0D4H,0D3H,0D1H,0D0H,0CEH,0CCH,0CBH,0C9H
        DB   0C7H,0C5H,0C3H,0C1H,0BFH,0BDH,0BBH,0B9H
        DB   0B7H,0B5H,0B3H,0B1H,0AEH,0ACH,0AAH,0A7H
        DB   0A5H,0A3H,0A0H,9EH,9BH,99H,97H,94H
        DB   92H,8FH,8DH,8AH,88H,85H,83H,80H
        DB   7DH,7BH,78H,76H,73H,71H,6EH,6CH
        DB   69H,67H,65H,62H,60H,5DH,5BH,59H
        DB   56H,54H,52H,4FH,4DH,4BH,49H,47H
        DB   45H,43H,41H,3FH,3DH,3BH,39H,37H
        DB   35H,34H,32H,30H,2FH,2DH,2CH,2AH
        DB   29H,28H,26H,25H,24H,23H,22H,21H
        DB   20H,1FH,1EH,1DH,1DH,1CH,1CH,1BH
        DB   1BH,1AH,1AH,1AH,19H,19H,19H,19H
        DB   19H,19H,19H,19H,1AH,1AH,1AH,1BH
        DB   1BH,1CH,1CH,1DH,1DH,1EH,1FH,20H
        DB   21H,22H,23H,24H,25H,26H,28H,29H
        DB   2AH,2CH,2DH,2FH,30H,32H,34H,35H
        DB   37H,39H,3BH,3DH,3FH,41H,43H,45H
        DB   47H,49H,4BH,4DH,4FH,52H,54H,56H
        DB   59H,5BH,5DH,60H,62H,65H,67H,69H
        DB   6CH,6EH,71H,73H,76H,78H,7BH,7DH
```
;三角波函数表
```
LIST1:  DB   80H,81H,82H,83H,84H,85H,86H,87H
        DB   88H,89H,8AH,8BH,8CH,8DH,8EH,8FH
        DB   90H,91H,92H,93H,94H,95H,96H,97H
        DB   98H,99H,9AH,9BH,9CH,9DH,9EH,9FH
        DB   0A0H,0A1H,0A2H,0A3H,0A4H,0A5H,0A6H,0A7H
        DB   0A8H,0A9H,0AAH,0ABH,0ACH,0ADH,0AEH,0AFH
```

```
        DB    0B0H,0B1H,0B2H,0B3H,0B4H,0B5H,0B6H,0B7H
        DB    0B8H,0B9H,0BAH,0BBH,0BCH,0BDH,0BEH,0BFH
        DB    0BFH,0BEH,0BDH,0BCH,0BBH,0BAH,0B9H,0B8H
        DB    0B7H,0B6H,0B5H,0B4H,0B3H,0B2H,0B1H,0B0H
        DB    0AFH,0AEH,0ADH,0ACH,0ABH,0AAH,0A9H,0A8H
        DB    0A7H,0A6H,0A5H,0A4H,0A3H,0A2H,0A1H,0A0H
        DB    9FH,9EH,9DH,9CH,9BH,9AH,99H,98H
        DB    97H,96H,95H,94H,93H,92H,91H,90H
        DB    8FH,8EH,8DH,8CH,8BH,8AH,89H,88H
        DB    87H,86H,85H,84H,83H,82H,81H,80H
        DB    7FH,7EH,7DH,7CH,7BH,7AH,79H,78H
        DB    77H,76H,75H,74H,73H,72H,71H,70H
        DB    6FH,6EH,6DH,6CH,6BH,6AH,69H,68H
        DB    66H,66H,65H,64H,63H,62H,61H,60H
        DB    5FH,5EH,5DH,5CH,5BH,5AH,59H,58H
        DB    55H,55H,55H,54H,53H,52H,51H,50H
        DB    4FH,4EH,4DH,4CH,4BH,4AH,49H,48H
        DB    44H,44H,45H,44H,43H,42H,41H,40H
        DB    40H,41H,42H,43H,44H,45H,46H,47H
        DB    48H,49H,4AH,4BH,4CH,4DH,4EH,4FH
        DB    50H,51H,52H,53H,55H,55H,56H,57H
        DB    58H,59H,5AH,5BH,5CH,5DH,5EH,5FH
        DB    60H,61H,62H,63H,66H,65H,66H,67H
        DB    68H,69H,6AH,6BH,6CH,6DH,6EH,6FH
        DB    70H,71H,72H,73H,77H,75H,76H,77H
        DB    78H,79H,7AH,7BH,7CH,7DH,7EH,7FH
  ;
        END                                          ;结束
```

5.6.2 C 源程序清单

```c
/*---------------------------------------
Wave Generator program V11.1
MCU STC89C52RC   XAL 12 MHz
Build by Gavin Hu, 2010.6.15
----------------------------------------*/
#include<reg51.h>
sbit UPKEY = P3^3;
sbit DOWNKEY = P3^4;
sbit SHKEY = P3^5;
sbit FOUT = P3^2;
void delay_ms(unsigned int);
```

```c
unsigned char fun = 0;
unsigned char th0_reg,tl0_reg;
/*------------------------------------
  main function
------------------------------------*/
void main(void)
{
unsigned int f = 500;
TMOD = 0x01;
th0_reg = (unsigned int)(65539.0-(19531.25/f))>>8;
tl0_reg = (unsigned int)(65539.0-(19531.25/f)) & 0x00ff;
TH0 = th0_reg;
TL0 = tl0_reg;
TR0 = 1;
IE = 0x82;
P3 = 0xff;
while(1)
    {
    if ((UPKEY==0)&&(f<1000))
        {
        f+=(f/10)? (f/10):1;
        th0_reg = (unsigned int)(65539.0-(19531.25/f))>>8;
        tl0_reg = (unsigned int)(65539.0-(19531.25/f)) & 0x00ff;
        delay_ms(100);
        }
    if ((DOWNKEY==0)&&(f>1))
        {
        f-=(f/10)? (f/10):1;
        th0_reg = (unsigned int)(65539.0-(19531.25/f))>>8;
        tl0_reg = (unsigned int)(65539.0-(19531.25/f)) & 0x00ff;
        delay_ms(100);
        }
    if (SHKEY==0)
        {
        fun = !fun;
        delay_ms(100);
        }
    }
}

/*------------------------------------
```

```c
    Delay function
    Parameter: unsigned int dt
    Delay time = dt(ms)
---------------------------------------*/
void delay_ms(unsigned int dt)
{
register unsigned char bt,ct;
for (; dt; dt--)
    for (ct=2;ct;ct--)
        for (bt=248; --bt; );
}


/*---------------------------------------
    T0 interrupt function
---------------------------------------*/
void intt0(void) interrupt 1
{
unsigned char code sin_table[] = {\
0,0,0,0,0,0,0,0,1,1,1,1,1,2,2,2,2,3,3,3,4,4,5,5,6,6,6,7,8,8,9,9,\
10,10,11,12,12,13,14,14,15,16,17,17,18,19,20,21,22,23,23,24,25,\
26,27,28,29,30,31,32,33,34,35,37,38,39,40,41,42,43,45,46,47,48,\
49,51,52,53,54,56,57,58,60,61,62,64,65,66,68,69,71,72,73,75,76,\
78,79,81,82,84,85,87,88,90,91,93,94,96,97,99,100,102,103,105,106,\
108,109,111,113,114,116,117,119,120,122,124,125,127,128,130,131,\
133,135,136,138,139,141,142,144,146,147,149,150,152,153,155,156,\
158,159,161,162,164,165,167,168,170,171,173,174,176,177,179,180,\
182,183,184,186,187,189,190,191,193,194,195,197,198,199,201,202,\
203,204,206,207,208,209,210,212,213,214,215,216,217,218,220,221,\
222,223,224,225,226,227,228,229,230,231,232,232,233,234,235,236,\
237,238,238,239,240,241,241,242,243,243,244,245,245,246,246,247,\
247,248,249,249,249,250,250,251,251,252,252,252,253,253,253,253,\
254,254,254,254,254,255,255,255,255,255,255,255,255};
static unsigned char wave_index = 0;
static bit index_sign = 1;
TL0 = tl0_reg;
TH0 = th0_reg;
if (fun) {P2 = sin_table[wave_index];}
    else {P2 = wave_index;}
FOUT = !FOUT;
if (index_sign == 1) {if (++wave_index == 255) index_sign = 0;}
    else {if (--wave_index == 0) index_sign = 1;}
}
```

第6章　16点阵LED显示器的设计

6.1　系统功能

设计一个能显示4个16×16点阵中文汉字或图形的LED显示屏,显示要求有静止、上移等显示方式。

6.2　设计方案

点阵LED图文显示的原理是,通过控制一些组成某些图形或文字的各个点所在位置相对应的LED器件是否发光,得到我们想要看到的显示结果。这种同时控制多个发光点亮灭的方法称为静态驱动显示方式。每个16×16的点阵文字共有256个发光二极管,显然单片机没有这么多端口,如果是4个文字,则需要有1 024个控制端口。如果采用锁存器来扩展端口,按8位的锁存器来计算,那么一个16×16的点阵就需要256/8=32个锁存器。在实际应用中显示屏往往还要大,锁存器的花费将是一个很庞大的数字。因此,在实际应用中显示屏采用另一种称为动态扫描的显示方法。

所谓动态扫描,就是逐行轮流点亮,这样扫描驱动电路就可以实现多行(比如16行)的同名列共用一套列驱动器。就16×16的点阵来说,我们把所有同一行的发光管的阳极连在一起,把所有同一列的发光管的阴极连在一起(共阳的接法),先送出对应第一行发光管亮灭的数据并锁存,然后选通第一行使其点亮一定的时间,然后熄灭;再送出第二行的数据并锁存,然后选通第二行使其点亮相同的时间,然后熄灭……第十六行之后又重新点亮第一行,这样反复轮回。当这样轮回的速度足够快(每秒24次以上)时,由于人眼的视觉暂留现象,我们就能看到显示屏上稳定的图形了。

采用扫描方式进行显示时,每行都有一个行驱动器,各行的同名列共用一个列驱动器。显示数据通常存储在单片机的存储器中,按8位一个字节的形式顺序排放。显示时要把一行中各列的数据都传送到相应的列驱动器上,这就存在一个显示数据传输的问题。从控制电路到列驱动器的数据传输,可以采用并行方式或串行方式。显然,采用并行方式时,从控制电路到列驱动器的线路数量大,相应的硬件数目多。当列数很多时,并行传输的方案是不可取的。

采用串行传输的方法时,控制电路可以只用一根信号线,将列数据一位一位传输到列驱动器,在硬件方面无疑是十分经济的。但是,串行传输过程较长,数据按顺序一位一位地传输出给列驱动器,只有当一行的各列数据都已传输到位之后,这一行的各列才能并行显示。这样,对于一行的显示过程就可以分解成列数据准备(传输)和列数据显示两个部分。对于串行传输方式来说,列数据准备时间相对要长一些,在行扫描周期确定的情况下,行显示的时间就会减小一点儿(比如4个水平排列的16×16文字的LED每行亮1 ms,则串行传送时间约为0.1 ms),水平文字较多时会影响到LED的亮度效果。

解决串行传输中列数据准备和列数据显示的时间矛盾问题时,可以采用重叠处理的方法,即在显示本行各列数据的同时,传送下一行的列数据。为了达到重叠处理的目的,列数据的显示就需要具有锁存功能。经过上述分析,可知列驱动器电路应具备能实现串入并出的移位功能,同时具有并行锁存输出的功能。这样,在本行已准备好的数据输入并行锁存器输出端口进行显示时,串并移位寄存器就可以准备下一行的列数据传送,而不会影响每行的显示时间。图6.1所示为16点阵LED显示器系统结构框图。

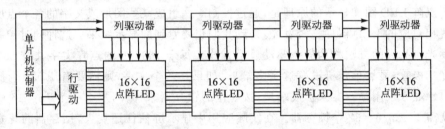

图6.1 16点阵LED显示器系统结构框图

6.3 系统硬件仿真电路的设计

系统硬件电路大致上可以分成单片机系统与外围电路、行驱动电路和列驱动电路3部分。

1. 单片机系统与外围电路

单片机采用AT89C52单片机,采用12 MHz或以上更高频率的晶振,以获得较高的刷新频率,使显示更稳定。单片机的串口与列驱动器相连,用来传送显示数据。P1口低4位与行驱动器相连(4/16译码器),送出行选信号;P1.5~P1.7口则用来发送移位控制信号。系统的Proteus硬件仿真电路如图6.2所示。

2. 行驱动电路

单片机P1口低4位输出的行号经4/16线译码器74LS154译码后生成16条行选通信号线,再经过驱动器驱动对应的行线。一条行线上要带动16列×4的LED进行显示,按每个LED器件需要5 mA电流计算,64个LED同时发光时,则需要320 mA电流,仿真电路中选用反相器并经74HC244驱动4个LED块的行扫描供电。

第6章 16点阵LED显示器的设计

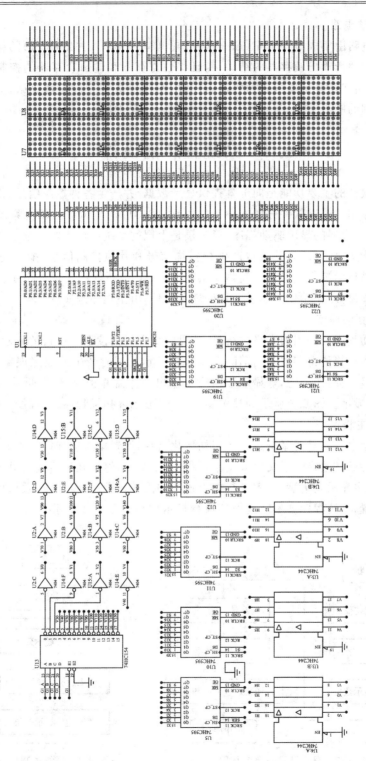

图6.2 系统的Proteus硬件仿真电路图

3. 列驱动电路

列驱动电路由集成电路 74HC595 构成,它具有一个串入并出(8 位)的移位寄存器和一个 8 位并行输出锁存器的结构,而且移位寄存器和输出锁存器的控制是各自独立的,可以实现在显示本行各列数据的同时,串行传送下一行的列数据,即具有时间上重叠处理的能力。

6.4 系统程序的设计

显示屏软件的主要功能是向屏幕提供显示数据,并产生各种控制信号,使屏幕按设计的要求显示。根据软件分层次设计的原理,我们可把显示屏的软件系统分成两大层:第一层是底层的显示驱动程序,第二层是上层的系统应用程序。显示驱动程序负责向屏体传送显示数据,并负责产生行扫描信号和其他控制信号,配合完成 LED 显示屏的扫描显示工作。显示驱动程序由定时器 T0 中断程序实现。系统应用程序完成系统环境设置(初始化)、显示效果处理等工作,由主程序来实现。

1. 显示驱动程序

显示驱动程序查询当前点亮的行号,从显示缓存区内读取下一行的显示数据,并通过串口发送给移位寄存器。为消除在切换行显示数据时产生的拖尾现象,驱动程序先要关闭显示屏,即消隐,等显示数据输入输出锁存器并锁存,然后再输出新的行号,重新打开显示。图 6.3 所示为显示驱动程序(显示屏扫描程序)流程图。

2. 系统主程序

系统主程序首先是对系统运行环境初始化,包括设置串口、定时器、中断和端口;然后以"静止翻屏"效果显示文字或图案,每次 4 个文字,停留几秒后接着显示后面的文字,一遍文字显示完毕后,接着进行向上连续的滚动显示汉字。显示效果可以根据需要进行设置,系统程序会不断地循环执行显示效果。图 6.4 所示是系统主程序的流程图。

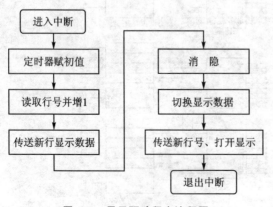

图 6.3 显示驱动程序流程图

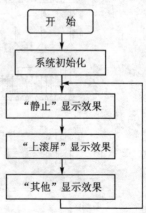

图 6.4 系统主程序流程图

6.5 软件调试与运行结果

LED 显示器的 Proteus 硬件仿真电路，只要器件连线可靠，一般无须调试即可正常工作。软件部分需要调试的主要有显示屏刷新频率及显示效果两部分。显示屏刷新频率由定时器 T0 的溢出率和单片机的晶振频率决定。从理论上来说，24 Hz 以上的刷新频率就能看到连续稳定的显示，刷新频率越高，显示越稳定；同时刷新频率越高，显示驱动程序占用 CPU 时间也越多。实验证明，在目测条件下，刷新频率为 40 Hz 以下的画面看起来闪烁较严重；刷新频率为 50 Hz 以上的已基本觉察不出画面闪烁；刷新频率达到 85 Hz 以上时，画面闪烁将没有明显改善。表 6.1 所列为采用 24 MHz 晶振时点阵显示屏刷新频率及其对应的定时器 T0 初值。

表 6.1　显示屏刷新频率(帧频)与 T0 初值关系表(24 MHz 晶振时)

刷新频率/Hz	25	50	62.5	75	85	100	120
T0 初值	0xEC78	0xF63C	0xF830	0xF97E	0xFA42	0xFB1E	0xFBEE

图 6.5 所示为用 Proteus 进行程序仿真运行的效果图。

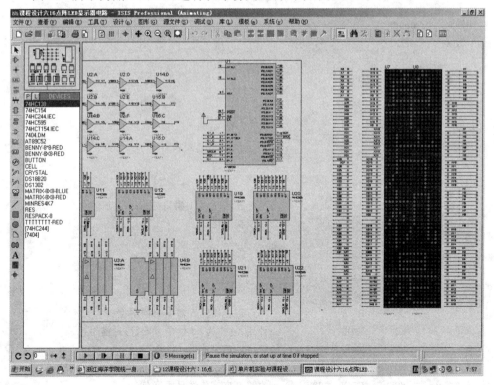

图 6.5　Proteus 程序仿真运行效果图

6.6 源程序清单

6.6.1 汇编源程序清单

```
;   *******************************
;   *          课程设计程序          *
;   *    4字16×16点阵电子屏字符显示器  *
;   *       AT89C52    12 MHz 晶振   *
;   *                                *
;   *******************************
;显示字用查表法,不占内存,用4个16×16共阳LED点阵显示
;效果:向上滚动显示字,每次4个,重复循环
;R2:行扫描地址(从00H~0FH)
;R3:滚动显示时控制移动速度,也可控制静止显示的时间
;***********;
;中断入口程序  ;
;***********;
;
        ORG     0000H
        LJMP    START
;
        ORG     000BH
        LJMP    INTT0
;
;***********;
;   主程序    ;
;***********;
;
START:  MOV     20H,#00H    ;清标志,00H为1帧扫描结束标志
        MOV     A,#0FFH     ;端口初始化
        MOV     P1,A
        MOV     P2,A
        MOV     P3,A
        MOV     P0,A
        CLR     P1.6        ;串行寄存器输入输出端控制位
        MOV     TMOD,#01H   ;使用T0作为16位定时器,行扫描用
        MOV     TH0,#0FCH   ;1 ms 初值(12 MHz)
        MOV     TL0,#18H
        MOV     SCON,#00H   ;串口0方式传送显示字节
        MOV     IE,#82H     ;T0中断允许,总中断允许
```

第 6 章 16 点阵 LED 显示器的设计

```
            MOV     SP,#70H
            LCALL   DIS1            ;显示准备,黑屏,1.5 s
MAIN:       MOV     DPTR,#TAB
            LCALL   MOVDISP         ;逐排显示,每次 4 字
            MOV     DPTR,#TAB
            LCALL   MOVDISP1        ;滚动显示,每排 4 字
            AJMP    MAIN
;
;
;*********************;
;4 字逐排显示子程序      ;
;*********************;
;每次 4 字移入移出显示方式,入口时定义 DPTR 值
;
MOVDISP:    MOV     R1,#6           ;显示 6 排字,每排 4 字(R1 = 排数)
DISLOOP:    MOV     R3,#100         ;每排显示时间 16 ms×100 = 1.6 s
DISMOV:     MOV     R2,#00H         ;第 0 行开始
            SETB    TR0             ;开扫描(每次一帧)
WAITMOV:    JBC     00H,DISMOV1     ;标志为 1 扫描一帧结束(16 ms 为 1 帧,每行 1 ms)
            AJMP    WAITMOV
DISMOV1:    DJNZ    R3,DISMOV       ;1 帧重复显示(控制显示时间)
            MOV     A,#128          ;显示字指针移一排(每排 4 字×32 = 128)
            ADD     A,DPL
            MOV     DPL,A
            MOV     A,#0
            ADDC    A,DPH
            MOV     DPH,A
            DEC     R1              ;R1 为 0,显示完
            MOV     A,R1
            JZ      MOVOUT
            AJMP    DISLOOP
MOVOUT:     RET                     ;移动显示结束
;
;**********************;
;4 字滚动显示子程序       ;
;**********************;
;每排 4 字向上移出显示方式,入口时定义 DPTR 值
;
MOVDISP1:   MOV     R1,#255         ;向上移动显示 6 排字,每排 4 字(R1 = 排数×16)
DISLOOP1:   MOV     R3,#10          ;移动速度 16 ms×10 = 0.16 s
DISMOV2:    MOV     R2,#00H         ;第 0 行开始
            SETB    TR0             ;开扫描(每次一帧)
```

```
WAITMOV1: JBC    00H,DISMOV3        ;标志为 1 扫描一帧结束(16 ms 为 1 帧,每行 1 ms)
          AJMP   WAITMOV1
DISMOV3:  DJNZ   R3,DISMOV2         ;1 帧重复显示(控制移动速度)
          INC    DPTR               ;显示字指针移一行(二字节位置)
          INC    DPTR
          DEC    R1
          MOV    A,R1               ;R1 为 0,显示完
          JZ     MOVOUT1
          AJMP   DISLOOP1
MOVOUT1:  RET                       ;移动显示结束
;
;*****************;
; 4 个字显示子程序      ;
;*****************;
;静止显示表中某 4 个字
DIS1:     MOV    R3,#5AH            ;静止显示时间控制
DIS11:    MOV    R2,#00H            ;一帧扫描初始值(行地址从 00H~0FH)
          MOV    DPTR,#TAB          ;取表首址
          SETB   TR0                ;开扫描(每次一帧)
WAIT11:   JBC    00H,DIS111         ;为 1,扫描一帧结束
          AJMP   WAIT11
DIS111:   DJNZ   R3,DIS11
          RET
;
;*************;
;  扫描程序       ;
;*************;
;1 ms 传送一行,每行显示 1 ms,一次传送 4 个字的某行共 8 个字节
;
INTT0:    PUSH   ACC
          MOV    TH0,#0FCH          ;1 ms 初值重装
          MOV    TL0,#18H
          MOV    A,#97              ;指向第 4 个字行右字节
          ADD    A,DPL
          MOV    DPL,A
          MOV    A,#0
          ADDC   A,DPH
          MOV    DPH,A
          MOV    A,#0
          MOVC   A,@A+DPTR          ;查表
          MOV    SBUF,A             ;串口 0 方式发送
WAIT:     JBC    TI,GO              ;等待发送完毕
```

第 6 章 16 点阵 LED 显示器的设计

```
            AJMP    WAIT
GO:         MOV     A,DPL           ;指向第 4 个字行左字节
            SUBB    A,#1
            MOV     DPL,A
            MOV     A,DPH
            SUBB    A,#0
            MOV     DPH,A
            MOV     A,#0
            MOVC    A,@A+DPTR
            MOV     SBUF,A
WAIT1:      JBC     TI,GO1
            AJMP    WAIT1
;
GO1:        MOV     R0,#03H
MLOOP:      MOV     A,DPL           ;指向前 3 个字行右字节
            SUBB    A,#31
            MOV     DPL,A
            MOV     A,DPH
            SUBB    A,#0
            MOV     DPH,A
            MOV     A,#0
            MOVC    A,@A+DPTR       ;查表
            MOV     SBUF,A          ;串口 0 方式发送
WAIT2:      JBC     TI,GO2          ;等待发送完毕
            AJMP    WAIT2
GO2:        MOV     A,DPL           ;指向前 3 个字行左字节
            SUBB    A,#1
            MOV     DPL,A
            MOV     A,DPH
            SUBB    A,#0
            MOV     DPH,A
            MOV     A,#0
            MOVC    A,@A+DPTR
            MOV     SBUF,A
WAIT3:      JBC     TI,GO3
            AJMP    WAIT3
GO3:        DJNZ    R0,MLOOP        ;执行 3 次
;
            SETB    P1.7            ;关行显示,准备刷新
            NOP                     ;串口寄存器数据稳定
            SETB    P1.6            ;产生上升沿,行数据输入输出端
            NOP
```

```
        NOP
        CLR     P1.6            ;恢复低电平
        MOV     A,R2            ;修改显示行地址
        ORL     A,#0F0H         ;修改显示行地址
        MOV     R2,A            ;修改显示行地址
        MOV     A,P1            ;修改显示行地址
        ORL     A,#0FH          ;修改显示行地址
        ANL     A,R2            ;修改显示行地址
        MOV     P1,A            ;修改完成
        CLR     P1.7            ;开行显示
        INC     R2              ;下一行扫描地址值
        INC     DPTR
        INC     DPTR            ;下一行数据地址
        MOV     A,R2
        ANL     A,#0FH
        JNZ     G04
        SETB    00H             ;R2 为 10H,现为末行扫描,置 1 帧结束标志
        MOV     A,DPL           ;指针修正为原帧初值
        SUBB    A,#32
        MOV     DPL,A
        MOV     A,DPH
        SUBB    A,#0
        MOV     DPH,A
        CLR     TR0             ;1 帧扫描完,关扫描
G04:    POP     ACC
        RETI                    ;退出
;
;***************;
;   扫描文字表        ;
;***************;
; 共 5 排字,每排 4 个字,前后为黑屏
TAB:
    DB  0FFH,0FFH,0FFH,0FFH,0FFH,0FFH,0FFH,0FFH,0FFH,0FFH,0FFH,0FFH,0FFH,
        0FFH    ;黑屏
    DB  0FFH,0FFH,0FFH,0FFH,0FFH,0FFH,0FFH,0FFH,0FFH,0FFH,0FFH,0FFH,0FFH,
        0FFH,0FFH
    DB  0FFH,0FFH,0FFH,0FFH,0FFH,0FFH,0FFH,0FFH,0FFH,0FFH,0FFH,0FFH,0FFH,
        0FFH    ;黑屏
    DB  0FFH,0FFH,0FFH,0FFH,0FFH,0FFH,0FFH,0FFH,0FFH,0FFH,0FFH,0FFH,0FFH,
        0FFH,0FFH
    DB  0FFH,0FFH,0FFH,0FFH,0FFH,0FFH,0FFH,0FFH,0FFH,0FFH,0FFH,0FFH,0FFH,
        0FFH    ;黑屏
```

第6章 16点阵LED显示器的设计

```
DB    0FFH,0FFH,0FFH,0FFH,0FFH,0FFH,0FFH,0FFH,0FFH,0FFH,0FFH,0FFH,0FFH,
      0FFH,0FFH
DB    0FFH,0FFH,0FFH,0FFH,0FFH,0FFH,0FFH,0FFH,0FFH,0FFH,0FFH,0FFH,0FFH,
      0FFH         ;黑屏
DB    0FFH,0FFH,0FFH,0FFH,0FFH,0FFH,0FFH,0FFH,0FFH,0FFH,0FFH,0FFH,0FFH,
      0FFH,0FFH
DB    0F7H,0DFH,0F9H,0CFH,0FBH,0BFH,0C0H,007H,0DEH,0F7H,0C0H,007H,0DEH,0F7H,0DEH,
      0F7H         ;单
DB    0C0H,007H,0DEH,0F7H,0FEH,0FFH,000H,001H,0FEH,0FFH,0FEH,0FFH,0FEH,0FFH,
      0FEH,0FFH
DB    0FFH,0BFH,0EFH,0BFH,0EFH,0BFH,0EFH,0BBH,0E0H,001H,0EFH,0FFH,0EFH,0EFH,
      0FFH         ;片
DB    0E0H,00FH,0EFH,0EFH,0EFH,0EFH,0EFH,0DFH,0EFH,0DFH,0EFH,0BFH,0EFH,
      07FH,0EFH
DB    0EFH,0FFH,0EFH,007H,0EFH,077H,001H,077H,0EFH,077H,0EFH,077H,0C7H,077H,0CBH,
      077H         ;机
DB    0ABH,077H,0AFH,077H,06EH,0F7H,0EEH,0F5H,0EDH,0F5H,0EDH,0F5H,0EBH,0F9H,
      0EFH,0FFH
DB    0FFH,0FFH,0F0H,00FH,0F7H,0EFH,0F0H,00FH,0F7H,0EFH,0F0H,00FH,0FFH,0FFH,000H,
      001H         ;是
DB    0FEH,0FFH,0F6H,0FFH,0F6H,007H,0F6H,0FFH,0EAH,0FFH,0DCH,0FFH,0BFH,001H,
      0FFH,0FFH
DB    0FFH,0FFH,0C0H,003H,0FEH,0FFH,0FEH,0FFH,0FEH,0FFH,0FEH,0FFH,0FEH,0FFH,0FEH,
      0FFH         ;工
DB    0FEH,0FFH,0FEH,0FFH,0FEH,0FFH,0FEH,0FFH,0FEH,0FFH,000H,001H,0FFH,0FFH,
      0FFH,0FFH
DB    0FBH,0BFH,0FBH,0BFH,0FBH,0BFH,0FBH,0BBH,0BBH,0B9H,0DBH,0B3H,0DBH,0B7H,0EBH,
      0AFH         ;业
DB    0E3H,0AFH,0EBH,09FH,0FBH,0BFH,0FBH,0BFH,0FBH,0BBH,000H,001H,0FFH,0FFH,
      0FFH,0FFH
DB    0FEH,0FFH,0FEH,0FFH,0DEH,0F7H,0C0H,003H,0DEH,0F7H,0DEH,0F7H,0DEH,0F7H,0DEH,
      0F7H         ;中
DB    0DEH,0F7H,0C0H,007H,0DEH,0F7H,0FEH,0FFH,0FEH,0FFH,0FEH,0FFH,0FEH,0FFH,
      0FEH,0FFH
DB    0E0H,00FH,0EFH,0EFH,0E0H,00FH,0EFH,0EFH,0E0H,00FH,0FFH,0FFH,000H,001H,0DDH,
      0FFH         ;最
DB    0C1H,003H,0DDH,077H,0C1H,0AFH,0DCH,0DFH,0C1H,0AFH,01DH,071H,0FCH,0FBH,
      0FDH,0FFH
DB                 ;
DB    0F7H,0DFH,0F7H,0DFH,080H,003H,0F7H,0DFH,0F0H,01FH,0F7H,0DFH,0F0H,01FH,0F7H,
      0DFH         ;基
DB    000H,001H,0F7H,0DFH,0EEH,0E7H,0C0H,011H,03EH,0FBH,0FEH,0FFH,080H,003H,
```

```
         0FFH,0FFH
    DB   0FEH,0FFH,0FEH,0FFH,0FEH,0FFH,000H,001H,0FCH,07FH,0FCH,0BFH,0FAH,0BFH,0FAH,
         0DFH    ;本
    DB   0F6H,0EFH,0EEH,0E7H,0D0H,011H,03EH,0FBH,0FEH,0FFH,0FEH,0FFH,0FEH,0FFH,
         0FFH,0FFH
    DB   0EFH,07FH,0EFH,07FH,0DFH,07FH,083H,003H,0BAH,0FBH,0BAH,0FBH,0B9H,0FBH,083H,
         07BH    ;的
    DB   0BBH,0BBH,0BBH,09BH,0BBH,0DBH,0BBH,0FBH,083H,0FBH,0BBH,0D7H,0BFH,0EFH,
         0FFH,0FFH
    DB   0FEH,0FFH,0FFH,07FH,0C0H,003H,0DFH,0FFH,0DDH,0FFH,0DEH,0F7H,0CFH,073H,0D7H,
         037H    ;应
    DB   0DBH,06FH,0DBH,06FH,0D9H,0DFH,0BBH,0DFH,0BFH,0BFH,0A0H,001H,07FH,0FFH,
         0FFH,0FFH
                 ;
    DB   0FFH,0FFH,0E0H,003H,0EFH,07BH,0EFH,07BH,0EFH,07BH,0E0H,003H,0EFH,07BH,0EFH,
         07BH    ;用
    DB   0EFH,07BH,0E0H,003H,0EFH,07BH,0EFH,07BH,0DFH,07BH,0DFH,07BH,0BFH,06BH,
         07FH,077H
    DB   0FDH,0FFH,0EFH,0FFH,0FFH,07FH,000H,001H,0FDH,0FFH,0FDH,0FFH,0FCH,00FH,0FDH,
         0EFH    ;方
    DB   0FBH,0EFH,0FBH,0EFH,0F7H,0EFH,0F7H,0EFH,0EFH,0DFH,06FH,03FH,09FH,
         0FFH,0FFH
    DB   0FFH,05FH,0FFH,067H,0FFH,06FH,080H,003H,0FFH,07FH,0FFH,07FH,0FFH,07FH,0C1H,
         07FH    ;式
    DB   0F7H,0BFH,0F7H,0BFH,0F7H,0BFH,0F4H,0DFH,0E3H,0DDH,08FH,0EDH,0DFH,0F5H,
         0FFH,0FBH
    DB   0F9H,0BFH,0C7H,0AFH,0F7H,0B7H,0F7H,0B7H,0F7H,0BFH,000H,001H,0F7H,0BFH,0F7H,
         0B7H    ;我
    DB   0F1H,0D7H,0C7H,0CFH,037H,0DFH,0F7H,0AFH,0F6H,06DH,0F7H,0F5H,0D7H,0F9H,
         0EFH,0FDH
    DB   0FFH,007H,0C0H,06FH,0EDH,0EFH,0F6H,0DFH,0C0H,001H,0DDH,0FDH,0BDH,0FFH,0C0H,
         003H    ;爱
    DB   0FBH,0FFH,0F8H,00FH,0F3H,0DFH,0F4H,0BFH,0EFH,03FH,09CH,0CFH,073H,0F1H,
         0CFH,0FBH
    DB   0F7H,0DFH,0F9H,0CFH,0BFH,0BFH,0C0H,007H,0DEH,0F7H,0C0H,007H,0DEH,0F7H,0DEH,
         0F7H    ;单
    DB   0C0H,007H,0DEH,0F7H,0FEH,0FFH,000H,001H,0FEH,0FFH,0FEH,0FFH,0FEH,0FFH,
         0FEH,0FFH
    DB   0FFH,0BFH,0EFH,0BFH,0EFH,0BFH,0EFH,0BBH,0E0H,001H,0EFH,0FFH,0EFH,0FFH,0EFH,
         0FFH    ;片
    DB   0E0H,00FH,0EFH,0EFH,0EFH,0EFH,0EFH,0EFH,0DFH,0EFH,0DFH,0EFH,0BFH,0EFH,
         07FH,0EFH
```

```
     DB     0EFH,0FFH,0EFH,007H,0EFH,077H,001H,077H,0EFH,077H,0EFH,077H,0C7H,077H,0CBH,
            077H      ;机
     DB     0ABH,077H,0AFH,077H,06EH,0F7H,0EEH,0F5H,0EDH,0F5H,0EDH,0F5H,0EBH,0F9H,
            0EFH,0FFH
                    ;
     END            ;结束
```

6.6.2 C 源程序清单

```
/ * --------------------------------------
16×64 点阵 LED 显示屏程序    V12.1
MCU AT89C52    XAL 24 MHz
Build by Gavin Hu, 2010.6.15
-------------------------------------- * /
#include "reg52.h"
#define BLKNUM 4
#define BREGNUM 32
sbit G = P1^7;                          //P1.7 为显示允许控制信号端口
sbit R_CLK = P1^6;                      //P1.6 为输出锁存器时钟信号端
sbit S_CLR = P1^5;                      //P1.5 为移位寄存器清零端
void delay_ms(unsigned int);            //延时函数
unsigned char idata dispram[BLKNUM][BREGNUM];  //显示区缓存
/ * --------------------------------------
   主函数 void main(void)
-------------------------------------- * /
void main(void)
{
#define CHARNUM 12
unsigned char code Bmp[][BREGNUM] = {\
/ * --  文字:  单   -- * /
/ * --  宋体 12;  此字体下对应的点阵为宽×高 = 16×16    -- * /
0xF7,0xDF,0xF9,0xCF,0xFB,0xBF,0xC0,0x07,0xDE,0xF7,0xC0,0x07,0xDE,0xF7,0xDE,0xF7,
0xC0,0x07,0xDE,0xF7,0xFE,0xFF,0x00,0x01,0xFE,0xFF,0xFE,0xFF,0xFE,0xFF,0xFE,0xFF,

/ * --  文字:  片   -- * /
/ * --  宋体 12;  此字体下对应的点阵为宽×高 = 16×16    -- * /
0xFF,0xBF,0xEF,0xBF,0xEF,0xBF,0xEF,0xBB,0xE0,0x01,0xEF,0xFF,0xEF,0xFF,0xEF,0xFF,
0xE0,0x0F,0xEF,0xEF,0xEF,0xEF,0xEF,0xDF,0xEF,0xBF,0xEF,0xBF,0xEF,0x7F,0xEF,

/ * --  文字:  机   -- * /
/ * --  宋体 12;  此字体下对应的点阵为宽×高 = 16×16    -- * /
0xEF,0xFF,0xEF,0x07,0xEF,0x77,0x01,0x77,0xEF,0x77,0xEF,0x77,0xC7,0x77,0xCB,0x77,
```

0xAB,0x77,0xAF,0x77,0x6E,0xF7,0xEE,0xF5,0xED,0xF5,0xED,0xF5,0xEB,0xF9,0xEF,0xFF,

/*-- 文字：文 --*/
/*-- 宋体12；此字体下对应的点阵为宽×高=16×16   --*/
0xFD,0xFF,0xFE,0xFF,0xFE,0xFF,0x00,0x01,0xF7,0xDF,0xF7,0xDF,0xF7,0xDF,0xFB,0xBF,
0xFB,0xBF,0xFD,0x7F,0xFE,0xFF,0xFD,0x7F,0xFB,0x9F,0xE7,0xE1,0x1F,0xF7,0xFF,0xFF,

/*-- 文字：本 --*/
/*-- 宋体12；此字体下对应的点阵为宽×高=16×16   --*/
0xFE,0xFF,0xFE,0xFF,0xFE,0xFF,0x00,0x01,0xFC,0x7F,0xFC,0xBF,0xFA,0xBF,0xFA,0xDF,
0xF6,0xEF,0xEE,0xE7,0xD0,0x11,0x3E,0xFB,0xFE,0xFF,0xFE,0xFF,0xFE,0xFF,0xFF,0xFF,

/*-- 文字：屏 --*/
/*-- 宋体12；此字体下对应的点阵为宽×高=16×16   --*/
0xC0,0x03,0xDF,0xFB,0xDF,0xFB,0xC0,0x03,0xDB,0xEF,0xDD,0xDF,0xD0,0x03,0xDD,0xDF,
0xDD,0xDF,0xC0,0x01,0xDD,0xDF,0xDD,0xDF,0xBB,0xDF,0xBB,0xDF,0x77,0xDF,0xEF,0xDF,

/*-- 文字：动 --*/
/*-- 宋体12；此字体下对应的点阵为宽×高=16×16   --*/
0xFF,0xDF,0xFF,0xDF,0x81,0xDF,0xFF,0xDF,0xFF,0x03,0x00,0xDB,0xEF,0xDB,0xEF,0xDB,
0xDB,0xDD,0xDB,0xB0,0xBB,0x05,0xBB,0xBF,0x7B,0xFE,0xEB,0xFD,0xF7,0xFF,0xFF,

/*-- 文字：态 --*/
/*-- 宋体12；此字体下对应的点阵为宽×高=16×16   --*/
0xFE,0xFF,0xFE,0xFF,0x80,0x03,0xFE,0xFF,0xFD,0x7F,0xFD,0xBF,0xFA,0xDF,0xF7,0x67,
0xCF,0xF9,0xFE,0xFF,0xFB,0x77,0xDB,0x7B,0xDB,0xED,0x9B,0xED,0xBC,0x0F,0xFF,0xFF,

/*-- 文字：效 --*/
/*-- 宋体12；此字体下对应的点阵为宽×高=16×16   --*/
0xEF,0xDF,0xF7,0xCF,0xF7,0xDF,0x80,0xDF,0xEB,0x81,0xED,0xBB,0xDE,0x3B,0xDD,0xBB,
0x9D,0xD7,0xEB,0xD7,0xF7,0xEF,0xF3,0xEF,0xED,0xD7,0xDF,0x31,0x3C,0xFB,0xFF,0xFF,

/*-- 文字：果 --*/
/*-- 宋体12；此字体下对应的点阵为宽×高=16×16   --*/
0xFF,0xFF,0xE0,0x0F,0xEE,0xEF,0xE0,0x0F,0xEE,0xEF,0xE0,0x0F,0xFE,0xFF,0xFE,0xFF,
0x00,0x01,0xFC,0x7F,0xFA,0xBF,0xF6,0xCF,0xCE,0xF1,0x3E,0xFB,0xFE,0xFF,0xFE,0xFF,

/*-- 文字：演 --*/
/*-- 宋体12；此字体下对应的点阵为宽×高=16×16   --*/
0xBF,0x7F,0xDF,0xBF,0xD8,0x01,0xFB,0xFB,0x7C,0x07,0xAF,0xBF,0xEC,0x07,0xDD,0xB7,
0xDC,0x07,0xDD,0xB7,0x3D,0xB7,0xBC,0x07,0xBF,0xFF,0xBE,0xEF,0xBE,0xF7,0xBD,0xF7,

第 6 章 16 点阵 LED 显示器的设计

```
/* -- 文字：示 -- */
/* -- 宋体12；此字体下对应的点阵为宽×高=16×16  -- */
0xFF,0xFF,0xE0,0x07,0xFF,0xFF,0xFF,0xFF,0xFF,0xFF,0x80,0x01,0xFE,0xFF,0xFE,0xFF,
0xEE,0xDF,0xEE,0xEF,0xDE,0xF7,0xBE,0xF3,0x7E,0xFB,0xFE,0xFF,0xFA,0xFF,0xFD,0xFF
};
unsigned char i,j,k,l;

P1 = 0xbf;                              //P1端口初值：允许接收、锁存、显示
SCON = 0x00;                            //串口工作模式0：移位寄存器方式
TMOD = 0x01;                            //定时器T0工作方式1：16位方式
for (i = 0;i < BLKNUM;i ++ )
    for (j = 0;j < BREGNUM;j ++ )
        dispram[i][j] = 0xff;
TR0 = 1;                                //启动定时器T0
IE = 0x82;                              //允许定时器T0中断
while (1)
    {
    for (i = 0;i < BLKNUM;i ++ )        //显示效果：卷帘入⌐
        for (j = 0;j < BREGNUM;j ++ )
            {
            dispram[i][j] = 0;
            if (j % 2) delay_ms(100);
            }                           //────────┘
    //delay_ms(2000);                   //延时2 s
    for (i = 0;i < BLKNUM;i ++ )        //显示效果：卷帘出⌐
        for (j = 0;j < BREGNUM;j ++ )
            {
            dispram[i][j] = Bmp[i][j];
            if (j % 2) delay_ms(100);
            }                           //────────┘
    for (l = BLKNUM;l < CHARNUM;l ++ )  //显示效果：上滚屏⌐
        for (k = 0;k < 32;k + = 2)
            {
            for (i = 0;i < (BLKNUM - 1);i ++ )
                {
                for (j = 0;j < (BREGNUM - 2);j ++ )
                    {
                    dispram[i][j] = dispram[i][j + 2];
                    }
                dispram[i][j] = dispram[i + 1][0];
                dispram[i][j + 1] = dispram[i + 1][1];
```

```c
            }
        for (j = 0;j<(BREGNUM - 2);j++)
            {
                dispram[i][j] = dispram[i][j + 2];
            }
        dispram[i][j] = Bmp[l][k];
        dispram[i][j + 1] = Bmp[l][k + 1];
        delay_ms(100);
        }                                       //————————
    delay_ms(3000);                             //延时 3 s
    }                                           //end while (1)
}


/* ----------------------------------------
   Delay function
   Parameter: unsigned int dt
   Delay time = dt(ms)
   ---------------------------------------- */
void delay_ms(unsigned int dt)
{
register unsigned char bt,ct;
for (; dt; dt--)
    for (ct = 2;ct;ct--)
        for (bt = 248; --bt; );
}

/* 显示屏扫描(定时器 T0 中断)函数 */
void leddisplay(void) interrupt 1 using 1
{
register unsigned char i,j = BLKNUM;
TH0 = 0xF8;                                     //设定显示屏刷新频率每秒 62.5 帧
TL0 = 0x30;
i = P1;                                         //读取当前显示的行号
i = ++i & 0x0f;                                 //行号加 1,屏蔽高 4 位
do {
   j--;
   TI = 0;
   SBUF = dispram[j][i*2+1];                    //送显示数据
   while (!TI);
   TI = 0;
   SBUF = dispram[j][i*2];                      //送显示数据
   while (!TI);
```

```c
    }while(j);            //完成一行数据的发送
G = 1;                    //消隐(关闭显示)
P1 &= 0xf0;               //行号端口清零
R_CLK = 1;                //显示数据输入输出锁存器
P1 |= i;                  //写入行号
R_CLK = 0;                //锁存显示数据
G = 0;                    //打开显示
}
```

第 7 章 电子密码锁的设计

7.1 系统功能

设计一个由单片机控制的电子密码锁,要求设定一组 6 位的数字开启密码并能够存储,在进行开启操作时,如果 3 次输入密码错误,则进行鸣叫报警并锁死系统;如果密码输入正确,则进行声光开启提示。

7.2 设计方案

本电子密码锁采用 C52 系列单片机作为控制器,显示器使用 8 个普通的七段共阳数码管,应用动态扫描完成信息显示;操作按键共设 12 个,除 0~9 这 10 个数字键外,还有 2 个密码设定与开锁操作键;报警与提醒电路采用扬声器与发光方式完成;存储器使用 AT24C01。总体的电路系统结构框图如图 7.1 所示。程序软件采用 Keil C51 编译器,用 C 语言编写控制代码。

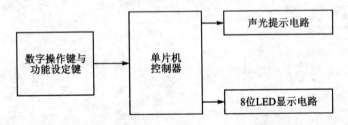

图 7.1 电子密码锁系统结构框图

7.3 系统硬件仿真电路的设计

电子密码锁系统硬件电路组成大致可以分成单片机控制电路、键盘电路、数码管显示电路、密码存储电路和声光提示电路 5 部分,图 7.2 所示为其硬件电路连接图。

1. 单片机控制电路

单片机端口资源的分配主要有:P0 口负责输出显示段码,端口接上拉电阻;P2 口为 LED 扫描控制口,对每个数码管进行约 1 ms 的轮流正电压供电;端口 P1.0~P1.4 分别对应数字键 0~4;端口 P3.0~P3.4 分别对应数字键 5~9;P3.5 端口为开启操作的功能键,开锁时先按一下开启操作键,然后再输入 6 位密码;P1.7 端口为修

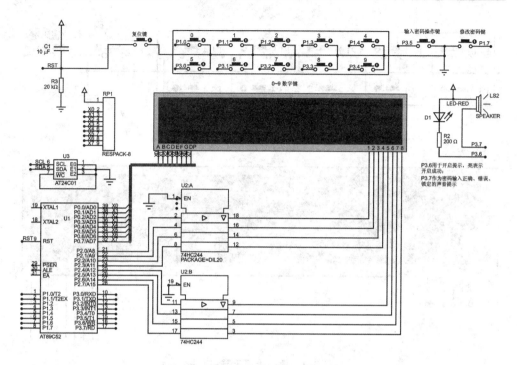

图 7.2　电子密码锁硬件电路连接图

改密码操作键,当需要更改密码时,先按一下修改密码键,然后输入 6 位的数字密码,当 6 位密码输入完毕时,系统会自动将新密码存入存储器;P3.6 端口控制一个发光二极管,用于锁开启提示,亮时表示开启成功,实际应用时可接电磁开锁线圈驱动电路;P3.7 端口连接一个扬声器作为密码输入正确或错误的声音提示,密码正确时响 3 次,前两次频率低,后一次频率高且时间较长;密码输入错误时响 3 次低频的声音,可再次输入密码,但总共只有 3 次机会,3 次密码错误后响 6 次较高频率的声音,然后自动锁死;P1.5 与 P1.6 端口分别为存储器 AT24C01 的时钟信号及数据信号端口。

2. 键盘电路

由于单片机端口数量足够,键盘电路直接使用了查询式按键电路,其中:小按键开关的一端接单片机端口,另一端接地,当按下按键时,通过单片机查询来确定哪个端口键按下了,然后执行相应的控制程序。图 7.3 所示为操作按键连接电路图,其中,2 个为功能键,10 个为 0~9 数字键。

3. 数码管显示电路

LED 显示器采用共阳八位一体的显示器 7SEG-MPX8-CA-BLUE,其中左边的 8 个引脚分别对应段码输出最低位至最高位,DP 为小数点位,而右边的 8 条引脚为八位数码管的阳极供电端。图 7.4 所示为数码管显示器引脚图。

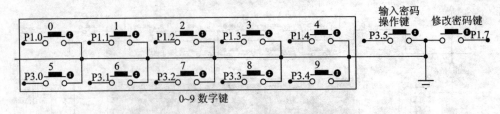

图 7.3　操作按键连接电路图

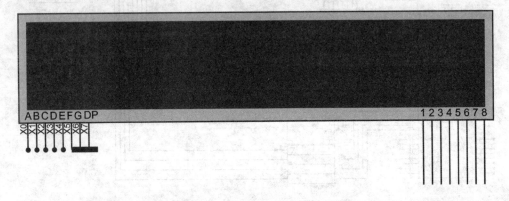

图 7.4　数码管显示器引脚图

4. 密码存储电路

存储器 AT24C01 是容量为 1 Kbit 的串行可电擦除只读存储器，可存储 128 个字节，该芯片支持 I^2C 总线数据传送协议。图 7.5 所示为 AT24C01 使用引脚连接图。

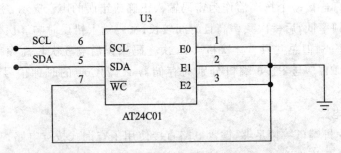

图 7.5　AT24C01 使用引脚连接图

5. 声光提示电路

声音与光提示电路使用了扬声器与红色发光二极管。扬声器用单片机输出的方波信号驱动发声，在密码输入错误或正确时使用不同频率的声音来区分。当红色发光二极管亮时，表示密码输入正确，开锁成功。图 7.6 所示为扬声器与红色发光二极管提示电路。

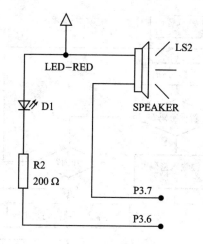

图 7.6　声光提示电路

7.4　系统程序的设计

单片机电子密码锁控制程序采用 C 语言编写,使用 Keil C51 编译器。控制程序模块主要有密码存取程序、开锁密码输入与比较程序、动态显示程序、发声程序等。

1. 主程序

主程序控制的主要流程如图 7.7 所示。开机上电后先进行一些初始化工作,然后等待开锁命令或设定新密码命令。当按下设定新密码键时,可输入 6 位的数字密码,系统在输入 6 位数字后自动将新密码存入存储器,并在下次开机时自动先取出存储的密码放入内存作为比较用。当按下开锁键时,系统在输入 6 位数字后自动进行密码比较,如果正确,则声光提醒并开锁;如果密码有误,则响 3 次警告声并进入继续输入密码状态,当输入密码的错误次数达到 3 次时,响 6 次报警声并锁住系统,此时不能进行任何操作。

2. 初始化程序

初始化程序主要是对单片机的控制寄存器进行一些工作前的设定,规划用于数据处理的内存,并将存储器中的密码读入内存用于开锁比较,调用显示程序使数码管全部发亮以检查发光二极管是否有损坏,最后进入主程序循环。

3. 按键扫描程序

按键扫描主要是先对开锁键及密码重新设定键进行检查,如果有按键按下,则进入相应的操作流程,此时等待读入数字键的键值并存入内存,当读入数字达到 6 个时,则进行存储或密码比较操作。

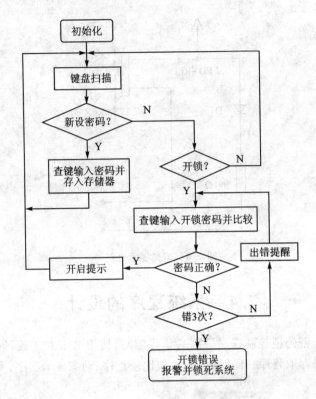

图 7.7 主程序工作流程图

7.5 软件调试与运行结果

软件编程调试按从小到大、从易到难的方法,首先在 Proteus 软件中打开 ISIS 软件,画出仿真电路图,然后使用编译器 Keil C51 进行程序的编写。主要编写与调试的模块程序的顺序如下:

① 扫描显示程序。扫描显示程序主要是对显示缓存中的 6 个数据进行 LED 段码扫描显示,调试内容包括共阳段码表的段码是否正确、每个 LED 是否能点亮、左边与右边的次序是否正确等。图 7.8 所示为检查段码是否能够全点亮的程序仿真运行图。

图 7.8 LED 显示 6 个"8"程序仿真图

② 提示信息显示程序。提示信息显示方式大致有输入密码提示信息、出错提示信息、开启成功提示信息等，图7.9～图7.14所示为各类信息显示图。

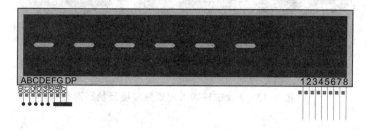

图7.9　等待输入新密码或等待输入开启密码状态图

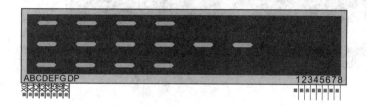

图7.10　输入新密码中状态图(图中已输入4位)

图7.11　输入开启密码状态图(图中已输入3位)

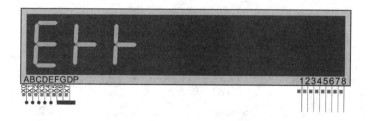

图7.12　输入密码出错提示状态图

③ 提示信息发声程序。提示发声程序的编程原理是，产生一定频率的方波信号输出给扬声器，电子密码锁程序中使用了3种提示发声，第一种是输入密码正确，为二低一高共3次的声响；第二种是输入密码错误，为3次低频声响；第三种是输入3次错误密码，为6次高频声响。

④ 存储器读/写程序。根据AT24C01存储器器件手册及I^2C通信标准编写存储器读/写程序，并将6个数据字节写入并读出进行比较，直到正确为止。

图 7.13 3 次密码输入出错报警提示状态图

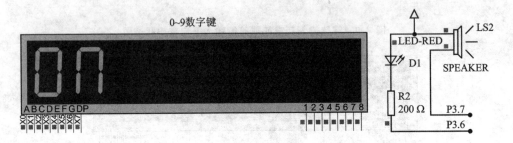

图 7.14 密码输入正确提示状态图(右边发光二极管亮表示开锁成功)

⑤ 读按键程序。按键程序采用顺序查键的方法,编程中应注意软件消抖以及按键等待释放等功能的使用。

⑥ 控制主程序。在以上各个子程序调试正确的基础上,可以按电子锁控制功能完成主程序的编写与调试。当然编程的方法也不是固定的,模块可以交叉编写或按自己的习惯编写。

7.6 源程序清单

```c
//*****************************************//
//              密码开锁演示程序               //
//                2013 年 1 月 18 日          //
//*****************************************//
//
//*************** 预处理 *******************//
#include <reg52.h>
#include "24c01.h"        //存储器 AT24C01 读/写程序
//
//*************** 定义 *********************//
sbit P10 = P1^0;
sbit P11 = P1^1;
sbit P12 = P1^2;
sbit P13 = P1^3;
```

第 7 章 电子密码锁的设计

```
sbit P14 = P1^4;
sbit P17 = P1^7;
sbit P30 = P3^0;
sbit P31 = P3^1;
sbit P32 = P3^2;
sbit P33 = P3^3;
sbit P34 = P3^4;
sbit P35 = P3^5;
sbit P36 = P3^6;
sbit P37 = P3^7;
//
char code dis_7[16] = {0xc0,0xf9,0xa4,0xb0,0x99,0x92,0x82,0xf8,0x80,0x90,0xbf,
                       0xc8,0xff,0xb6,0x86,0x8f};
// 共阳显示段码表:      0    1    2    3    4    5    6    7    8    9
//                      -    n    不亮  三    E    r
char code scan_con[6] = {0x01,0x02,0x04,0x08,0x10,0x20};
//控制七段数码管的工作电压,0x01 表示最左端数码显示
char data dia[6] = {0x08,0x08,0x08,0x08,0x08,0x08};
//存储显示于数码管上的值,最开始显示"888888",顺便检验数码管有没有损坏的现象
unsigned char data dat1[2][6] = {0,0,0,0,0,0};
//用于存储设定密码与输入密码数值;dat1[0][]为设定的密码,dat1[1][]为输入的密码
unsigned char data dat2[6];//显示缓存
bit flag = 0,flag1 = 0;
//
// ****************** 函数声明 ********************//
//
void delayms(int t);
void scan(void);
void sound(char a, char b,char c);
void input(char x);
void displayon(void);
void displaywrong(void);
void displayagain(void);
void displayinput(void);
void jiemi(void);
// ****************** 1 ms 延时程序 ********************//
//
void delayms(int t)
{
    int i, j;
    for(i = 0;i < t;i ++ )
        for(j = 0;j < 120;j ++ )
```

```c
        ;
    }
//***************显示扫描程序*************************//
//
void scan(void)
{
    unsigned char k;
    for(k=0;k<6;k++)
    {
        P0 = dis_7[dia[k]];
        P2 = scan_con[k];
        delayms(1);
        P0 = 0xff;
    }
}
//
//****************发声程序*********************//
//a表示响的时间,b表示频率,c表示响的次数
void  sound(char a, char b,char c)
        {
            char i, j,k;
            for(k=0;k<c;k++)
            {
                for(i=0;i<a;i++)
                    for(j=0;j<100;j++)
                    {
                        P37 = !P37;
                        delayms(b);
                    }
                delayms(700);
            }
        }
//
//************按键输入处理程序*********************//
//
void input(char x)          //参数x=0表示输入设定值;x=1表示输入密码
{
    char i, j=0;            //j作为是否有按键输入的标志位
    for(i=0;i<6;i++ ,j=0)
        while(j==0)
        {
```

```
      scan();
      if(P10 == 0)                    //10 端口表示设定 0
      {
         while(P10 == 0)    scan();
         if(x == 0)
            dia[i] = 13;
         else
            dia[i] = 0;
         j = 1;
         dat1[x][i] = 0;
      }
      if(P11 == 0)                    //11 端口表示设定 1
      {
         while(P11 == 0)    scan();
         if(x == 0)
            dia[i] = 13;
         else
            dia[i] = 1;
         j = 1;
         dat1[x][i] = 1;
      }
      if(P12 == 0)                    //12 端口表示设定 2
      {
         while(P12 == 0)    scan();
         if(x == 0)
            dia[i] = 13;
         else
            dia[i] = 2;
         j = 1;
         dat1[x][i] = 2;
      }
      if(P13 == 0)                    //13 端口表示设定 3
      {
         while(P13 == 0)    scan();
         if(x == 0)
            dia[i] = 13;
         else
            dia[i] = 3;
         j = 1;
         dat1[x][i] = 3;
      }
      if(P14 == 0)                    //14 端口表示设定 4
```

```
    {
        while(P14 == 0)    scan();
        if(x == 0)
            dia[i] = 13;
        else
            dia[i] = 4;
        j = 1;
        dat1[x][i] = 4;
    }
    if(P30 == 0)                    //30 端口表示设定 5
    {
        while(P30 == 0)    scan();
        if(x == 0)
            dia[i] = 13;
        else
            dia[i] = 5;
        j = 1;
        dat1[x][i] = 5;
    }
    if(P31 == 0)                    //31 端口表示设定 6
    {
        while(P31 == 0)    scan();
        if(x == 0)
            dia[i] = 13;
        else
            dia[i] = 6;
        j = 1;
        dat1[x][i] = 6;
    }
    if(P32 == 0)                    //32 端口表示设定 7
    {
        while(P32 == 0)    scan();
        if(x == 0)
            dia[i] = 13;
        else
            dia[i] = 7;
        j = 1;
        dat1[x][i] = 7;
    }
    if(P33 == 0)                    //33 端口表示设定 8
    {
        while(P33 == 0)    scan();
```

```c
            if(x == 0)
               dia[i] = 13;
            else
               dia[i] = 8;
            j = 1;
            dat1[x][i] = 8;
         }
         if(P34 == 0)                          //34端口表示设定9
         {
            while(P34 == 0)    scan();
            if(x == 0)
               dia[i] = 13;
            else
               dia[i] = 9;
            j = 1;
            dat1[x][i] = 9;
         }
      }
   }
}
//***************** 开锁成功显示信息 *******************//
//
void displayon(void)                  //当密码正确时显示on
{
   dia[0] = 0x00; dia[1] = 0x0b;
   dia[2] = 0x0c; dia[3] = 0x0c; dia[4] = 0x0c; dia[5] = 0x0c;
   scan();
}
//*************** 密码输入3次错误后显示信息 ****************//
//当显示"------"时等待输入密码,当输入3次密码仍然错误时显示ErrErr
//
void displaywrong(void)
{
   dia[0] = 0x0e; dia[1] = 0x0f; dia[2] = 0x0f;
   dia[3] = 0x0e; dia[4] = 0x0f; dia[5] = 0x0f;
   scan();
}
//*************** 密码输入提示显示信息 ******************//
//当输入设定值之后显示"------",等待输入密码显示"------"
//
void displayinput(void)
{
   dia[0] = 0x0a; dia[1] = 0x0a; dia[2] = 0x0a;
```

```
    dia[3] = 0x0a; dia[4] = 0x0a; dia[5] = 0x0a;
    scan();
}
//**************密码出错再输入提示程序******************//
//当密码输入错误,要求再次输入时,先显示 Err,3 s 后显示"------"
//
void displayagain(void)
{
    int n;
    dia[0] = 0x0e; dia[1] = 0x0f; dia[2] = 0x0f;
    dia[3] = 0x0c; dia[4] = 0x0c; dia[5] = 0x0c;
//  以下显示 Err 约 3 s
    for(n = 0;n<500;n++) scan();
//  以下显示"------"
    dia[0] = 0x0a; dia[1] = 0x0a; dia[2] = 0x0a;
    dia[3] = 0x0a; dia[4] = 0x0a; dia[5] = 0x0a;
    scan();
}
//*****************密码核对程序***********************//
//
void jiemi(void)
    {
        char i, j, m = 0;
        for(i = 0;i<3;i++)
        {
            input(1);
            for(j = 0;j<6;j++)
            {
              if(dat1[0][j]! = dat1[1][j])
                {
                    m = 1;
                    break;
                }
            }
            if(m == 1)
            {
                if(i<2)
                {
                    sound(3,3,3);    //a 表示时间,b 表示频率,人耳辨别范围为 20~20 kHz
                                     //c 表示响的次数
                    displayagain();
                    m = 0;
```

```
            }
            else
            {
                sound(6,1,6);
                flag = 1;
                displaywrong();
            }
        }
        else if(m == 0)
        {
            sound(2,2,2);sound(5,1,1);
            P36 = 0;
            displayon();
            flag1 = 1;
            break;
        }
    }
}
//
//*************** 主程序 ********************//
//
void main(void)
{
    //input(0);
    //delayms(1000);
    // jiemi();
    //sound(2,2,1);                    //b 越大,频率越低
    //sound(5,1,1);
    char i;
    init();                            //初始化 AT24C01
    for(i = 0;i<6;i++)
    {
    //write_add(00+i,dat1[0][i]);      //在 23 地址处写入数据 0x55
    //delay1(100);
        dat1[0][i] = read_add(00+i);   //读入 AT24C01 存储器中的密码数据(共 6 个)
        delay1(100);
    }
//
    while(1)
    {
        scan();                        //显示扫描程序
        if(P35 == 0)
```

```c
            {
                if(flag == 0)
                {
                    P36 = 1;
                    displayinput();
                    jiemi();
                }
            }
        if(P17 == 0)
        {
            while(P17 == 0) scan();
            if(flag == 0)
            {
                displayinput();                          //显示"------"等待输入密码
                input(0);
                for(i = 0;i<6;i++)
                {
                    write_add(00 + i,dat1[0][i]);       //在 23 地址处写入数据 0x55
                    delay1(100);
                    //dat1[0][i] = read_add(00 + i);  用于测试写入与读出的数据是否一样
                    //delay1(100);
                }
                if(flag1 == 1)
                    displayon();
                else
                    displayinput();
            }
        }
    }
}
//***************** 程序结束 *********************//
```

以下为 AT24C01 头文件：

```
//*******************************************************//
//           AT24C01.H 头文件                             //
//              2013 年 1 月 18 日                        //
//*******************************************************//
//
#ifndef _24c01_H_
#define _24c01_H_
//
//***************** 预处理 *********************//
```

```c
#define uchar unsigned char
//****************定义********************//
sbit sda = P1^6;
sbit scl = P1^5;
uchar a;
//****************延时程序******************//
void delay()
{ ;; }
//*******************************************//
//开始信号
void start()
{
    sda = 1;
    delay();
    scl = 1;
    delay();
    sda = 0;
    delay();
}
//*******************************************//
//停止
void stop()
{
    sda = 0;
    delay();
    scl = 1;
    delay();
    sda = 1;
    delay();
}
//*******************************************//
//应答,在数据传送8位后,等待或者发送一个应答信号
void respons()
{
    uchar i;
    scl = 1;
    delay();
    while((sda == 1)&&(i<250))i++;
    scl = 0;
    delay();
}
//*******************************************//
```

```c
//初始化函数,拉高 sda 和 scl 两条总线
void init()
{
    sda = 1;
    scl = 1;
}
/***************************************************************/
/*根据数据有效性规则,读写数据时必须将 scl 拉高,然后送入或读出数据,完毕后再
  将 scl 拉低*/
/***************************************************************/
//写一字节,将 date 写入 AT24C01 中
void write_byte(uchar date)
{
    uchar i;
    scl = 0;
    for(i = 0;i < 8;i ++ )
    {
        date = date<<1;
        sda = CY;                   //将要送入的数据送入 sda
        scl = 1;                    //scl 拉高准备写数据
        delay();
        scl = 0;                    //scl 拉低数据写完毕
        delay();
    }
}
uchar read_byte()                   //读取一字节,从 AT24C01 中读取一个字节
{
    uchar i,k;
    for(i = 0;i < 8;i ++ )
    {
        scl = 1;                    //scl 拉高准备读数据
        delay();
        k = (k<<1)|sda;             //将 sda 中的数据读出
        scl = 0;                    //scl 拉低数据写完毕
        delay();
    }
    return k;
}
//*****************************************************************//
//延时程序
void delay1(uchar x)
```

```c
{
    uchar a,b;
    for(a = x;a>0;a--)
    for(b = 100;b>0;b--);
}
/******************************************************/
/* 读出与写入数据时必须严格遵守时序要求 */
/******************************************************/
//向 AT24C01 中写数据
void write_add(uchar address,uchar date)
{
    start();
    write_byte(0xa0);
    respons();
    write_byte(address);
    respons();
    write_byte(date);
    respons();
    stop();
}
//****************************************************//
//从 AT24C01 中读出数据
uchar read_add(uchar address)
{
    uchar date;
    start();
    write_byte(0xa0);
    respons();
    write_byte(address);
    respons();
    start();
    write_byte(0xa1);
    respons();
    date = read_byte();
    stop();
    return date;
}
#endif
//
//***************** AT24C01 头文件结束 *****************//
```

第 8 章　ISD4002 语音录放电路的设计

8.1　系统功能

利用单片机及语音录放芯片制作一个可录放语音的电路系统,可用于汽车站点播报、电话语音自动转接留言、银行 ATM 取款服务等语音信息自动提示场合。

8.2　设计方案

声音的记录从最早的模拟方式发展到现在的数字化方式,处理技术更加先进、可靠。一般的声音数字化记录过程是先经过模/数(A/D)转换,并按一定时间间隔的取样数字化值存在 EPROM 存储器中,放音时再按一定的时间间隔将存储器的数据进行数/模(D/A)转换,并进行滤波、音频放大等过程,还原记录的声音信息。由于受存储器容量及采样速度的限制,音质与成本会有一定的矛盾,对普通电子爱好者来说制作难度也较大。

ISD4002 是美国 ISD 公司系列单片语音录放集成电路系列中的一种,它采用直接模拟量存储技术,将每个采样值直接存储在片内的 Flash RAM 中,能较好地保留模拟量中的有效成分。芯片设计成和微控制器配合使用的寻址和控制方式,使器件引出端数目减到最少,且音质较好,目前在语音合成设计中应用较广。本设计采用 ISD4002 - 120P 芯片,可录音时间为 120 s,芯片供电电源为 3 V,工作电流为 15～20 mA,典型待机维持电流为 1 μA,可实现循环多段语音录放编程使用,芯片内部采样频率为 8 kHz,音质较好,适于一般电话以及其他语音提示设备的应用。图 8.1 所示为 ISD4002 语音录放系统构成框图。

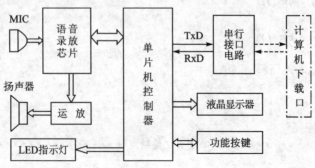

图 8.1　ISD4002 语音录放系统构成框图

8.3 系统硬件电路的设计

语音录放系统的电路原理图如图 8.2 所示。语音录放电路控制器采用宏晶公司的 STC89C52RC 单片机，它是一款低功耗、低电压、高性能的 8 位单片机，除兼容 51 系列单片机外，内部还具有 2 KB 的 EEPROM 存储器，方便保存历史数据；另外，它还具备 ISP 在线下载程序功能，方便产品开发阶段的程序调试。ISD4002-120P 录放芯片接收单片机的指令，执行录音或放音操作；LCD 中文液晶显示器对电路执行过程进行信息提示；按键开关采用顺序查询式读入单片机，并按要求执行相应功能；电源电路提供 5 V 及 3.3 V 两组电压给单片机及语音录放芯片。主要电路原理如下：

1. 电源电路

电源电路根据使用场合可选择交流输入或电池组模式，交流输入模式下采用降压、整流、滤波、稳压等，产生电路所需的各档电压。由于系统电路板耗电较少，故设计中采用 LM7805、AMS1117-3.3 三端稳压集成块，分别输出 5 V 及 3.3 V 的工作电压。

2. 语音信号拾取

语音信号拾取采用小体积的柱极体电容话筒，前置放大采用三极管 9014，ISD4002 声音输入端在单端输入时一般信号幅度不超过 32 mV。如果电容话筒的灵敏度较高，则前置放大级也可取消。本设计方案采用单端输入方式。

3. 音频放大电路

音频放大电路采用运放电路 LM386。该电路外围电路极其简单，按图 8.2 中连接，放大倍数为 200 倍左右，在 8 Ω 的扬声器上具有足够的音频功率，输入端的电位器可调整放音时的音量大小。

4. 显示电路

为方便进行录音、放音时的操作，显示系统中使用了中文液晶显示器 12232F。该显示器具有两行中文（共 14 个汉字）或两行 ASCII 字符（30 个）显示功能，并具有并口或串口通信功能。设计中采用了串口通信，这样可以节约单片机的端口资源。其中，P0.0 为液晶背光灯控制引脚，P0.1 为片选控制引脚，P0.2 为数据引脚，P0.3 为同步时钟信号引脚。另外，在 P0.5 及 P0.6 引脚接了 2 个发光二极管，用于录音及放音时的指示。

5. 串行通信电路

宏晶单片机具有 ISP 在线下载程序的功能，因此电路中设计了 MAX232 芯片组成的 RS-232 格式串行口通信电路，在调试程序时可随时下载程序进行实际运行观察，以便及时发现问题。

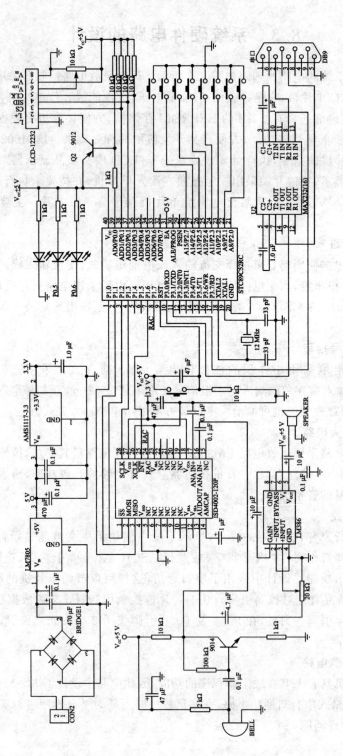

图 8.2 语音录放系统的电路原理图

第8章 ISD4002语音录放电路的设计

6. 按键电路

电路板中设置了8个按键,接在单片机的P2口。由于语音录放电路操作简单,实际仅使用了2个设定键,用于录音及放音控制,另6个按键备用。

7. 语音录放电路

ISD4002-120可录放120 s语音信号,具有600行使用地址,每5行地址为1 s,每一行均可寻址,其地址范围为000H~258H(即十进制数0~600)。系统设计中分成2段录音,第1段地址从000H开始,第2段地址从12CH(即十进制数300)开始,具体应用中可根据需要进行分段及相应的地址计算。表8.1所列为ISD4002操作指令表,详细资料请参考相关产品说明书。

表 8.1 ISD4002 操作指令表

指 令	5位控制码<11位地址>	操作摘要
POWERUP	00100<××××××××××× >	上电。等待TPUD后器件可以工作
SET PLAY	11100< A10~A0>	从指定地址开始放音。必须后跟PLAY指令使放音继续
PLAY	11110<××××××××××× >	从当前地址开始放音(直至EOM或OVF)
SET REC	10100<A10~A0>	从指定地址开始录音。必须后跟REC指令使录音继续
REC	10110<××××××××××× >	从当前地址开始录音(直至OVF或停止)
SET MC	11101<A10~A0>	从指定地址开始快进。必须后跟MC指令使快进继续
MC	11111<××××××××××× >	执行快进,直到EOM。若再无信息,则进入OVF状态
STOP	0×110<××××××××××>	停止当前操作
STOP WRDN	0×01×<××××××××××>	停止当前操作并掉电
RINT	0×110<××××××××××>	读状态:OVF和EOM

8.4 系统程序的设计

语音录放电路控制程序主要包含录音程序、放音程序、主循环程序等。录音程序受录音按键控制,放音程序受放音按键控制,在主程序循环中来回检测录、放按键,当检测到有按键按下时执行相应的功能。录音程序和放音程序的流程图如图8.3和图8.4所示。

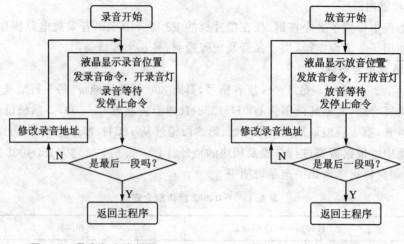

图 8.3　录音程序流程图　　　　图 8.4　放音程序流程图

8.5　调试及性能分析

系统调试可分为两部分:硬件电路调试先在不通电的情况下进行电路板的静态测试,确认无误后再通电测试,接着进行程序的编写及运行测试;程序先调试 LCD 显示器功能,然后调试录音及放音程序。在录音过程中,可以通过示波器测试语音芯片的 RAC 端,正常状态下,会有 175 ms 高电平、25 ms 低电平的周期信号,若能检测到,则说明芯片已经开始工作。

程序调试完成后,系统能进行两段各 60 s 长的语音录放,并且系统运行稳定。若要进行多段录、放音,则只须更改程序中的段数及每段录音时间定义即可。

8.6　源程序清单

```
/*-----------------------------------------------------
              ISD4002-120 record/play program
              MCU STC89C52RC   XTAL 12 MHz
              Display   12232F
-----------------------------------------------------*/
//以下程序能录两段声音,每段 60 s,采用顺序录音与顺序放音
//#pragma   src
#include <reg52.h>
//
#define uchar unsigned char
#define uint unsigned int
```

第 8 章 ISD4002 语音录放电路的设计

```c
#define ulong unsigned long
//
/*--------------------------------------------
                    端口定义
----------------------------------------------*/
//
sbit LED = P0^5;                    //录音指示,绿色
sbit LED1 = P0^6;                   //放音指示,黄色
sbit BUTTON1 = P2^0;                //录音开关
sbit BUTTON2 = P2^1;                //放音开关
sbit BUTTON3 = P2^2;                //备用
sbit BUTTON4 = P2^3;                //备用
sbit BUTTON5 = P2^4;                //备用
sbit BUTTON6 = P2^5;                //备用
sbit BUTTON7 = P2^6;                //备用
sbit BUTTON8 = P2^7;                //备用
//
//ISD4002 控制口
sbit ISD4002_INT = P1^6;            //ISD4002 中断输出
sbit ISD4002_RAC = P1^7;            //ISD4002 行地址时钟输出
sbit ISD4002_SS = P1^2;             //ISD4002 器件选择
sbit SCK = P1^3;                    //ISD4002 串行时钟,由单片机提供
sbit MOSI = P1^1;                   //ISD4002 串行输入端
sbit MISO = P1^0;                   //ISD4002 串行输出端
//
//12232F 液晶接口
sbit    BLACK = P0^0;               //背光灯
sbit    CS = P0^1;                  //使能口
sbit    SID = P0^2;                 //数据口
sbit    SCLK = P0^3;                //时钟口
//
//显示数据区定义
uchar   code    DISP_TAB[10] = {"0123456789"};              //显示查表用
uchar   code    REC_END[15] = {"------录音结束"};            //信息显示表
uchar   code    REC_CON[15] = {"----请继续录音"};            //信息显示表
uchar   code    PLAY_END[15] = {"------放音结束"};           //信息显示表
uchar   code    PLAY_CON[15] = {"----请继续放音"};           //信息显示表
uchar   code    LOG_DATA[15] = {" -电子式录音机-"};          //信息显示表
uchar   xdata   REC_DISP_DATA[15] = {"第 00 段录音中  "};    //信息显示表
uchar   xdata   PLAY_DISP_DATA[15] = {"第 00 段放音中  "};   //信息显示表
/*--------------------------------------------
                    功能函数
```

```c
//以下为 12232F 液晶程序
/************1 ms 延时函数 **************/
delay1ms(uint t)
{
int i,j;
for(i = 0;i<t;i++)
    for(j = 0;j<120;j++)                                    //12 MHz 晶振时
    ;
}
//
/**********向 12232F 写 1 字节 ************/
void write_data(uchar dat_d)
{
    uchar i,val;
    delay1ms(1);CS = 1;val = 0xFA;
    for (i = 8; i>0; i--)
    {
    SID = val>>7; SCLK = 0;SCLK = 1;val = val<<1;}
    val = dat_d&0xF0;
    for (i = 8; i>0; i--)
    {
    SID = val>>7; SCLK = 0;SCLK = 1;val = val<<1;}
    val = dat_d&0x0F;
    val = val<<4;
    for (i = 8; i>0; i--)
    {
    SID = val>>7; SCLK = 0;SCLK = 1;val = val<<1;}
    CS = 0;
}
//
/*********向 12232F 写 1 个命令字节 ************/
void write_com(uchar com_d)
{
    uchar i,val;
    delay1ms(1);CS = 1;val = 0xF8;
    for (i = 8; i>0; i--)
    {
    SID = val>>7; SCLK = 0;SCLK = 1;val = val<<1;}
    val = com_d&0xF0;
    for (i = 8; i>0; i--)
```

```c
        {
        SID = val≫7; SCLK = 0;SCLK = 1;val = val≪1;}
        val = com_d&0x0F;
        val = val≪4;
        for (i = 8; i>0; i--)
        {
        SID = val≫7; SCLK = 0;SCLK = 1;val = val≪1;}
        CS = 0;
}
/**********12232F 初始化**************/
void   setup()
{
    CS = 0;                            //以下为 12232F 初始化
    delay1ms(400);                     //延时 400 ms 以上
    write_com(0x01);
    write_com(0x30);
    write_com(0x02);
    write_com(0x04);
    write_com(0x0C);
    write_com(0x01);
    write_com(0x80);
    delay1ms(400);                     //延时 400 ms 以上
    //
}
//
//*********** 液晶写入函数 ***********//
// _wz 为 0x80(液晶第 1 行)或 0x90(液晶第 2 行)
// _len 为 0～15(每行字符个数)
// _adr 为数据表名
void _write(uchar _wz,uchar _len,uchar * _adr)
{
    uchar m;
    //q = _wz;
    write_com(_wz);                    //写入字定位(起始位)
    for (m = 0; m<_len; m ++ )
    {write_data( * (_adr + m));}
    }
    //12232F 液晶程序结束
    //

/* -------------------------------------------------
```

```
                        SPI 读/写函数
------------------------------------------------*/
uchar SPI_rw(uchar dat)
{
    uchar i;
    for (i = 0; i < 8; i++)
        {
            MOSI = dat & 0x01;
            SCK = 1;
            dat >>= 1;
            dat |= (uchar)MISO << 7;
            SCK = 0;
        }
    return dat;
}


/*------------------------------------------------
                     ISD4002 电源唤醒
------------------------------------------------*/
void ISD4002_powerup(void)
{
    ISD4002_SS = 0;
    SPI_rw(0x20);
    ISD4002_SS = 1;
}


/*------------------------------------------------
                     ISD4002 停止命令
------------------------------------------------*/
void ISD4002_stop(void)
{
    ISD4002_SS = 0;
    SPI_rw(0x30);
    ISD4002_SS = 1;
}


/*------------------------------------------------
                   ISD4002 按地址录音
------------------------------------------------*/
void ISD4002_record(uint addr)
{
    ISD4002_SS = 0;
```

```c
    SPI_rw((uchar)(addr & 0x00FF));
    SPI_rw(((uchar)(addr>>8) & 0x03) | 0xA0);
    ISD4002_SS = 1;
}


/* ---------------------------------------------
                    ISD4002 按地址放音
---------------------------------------------- */
void ISD4002_play(uint addr)
{
    ISD4002_SS = 0;
    SPI_rw((uchar)(addr & 0x00FF));
    SPI_rw(((uchar)(addr>>8) & 0x03) | 0xE0);
    ISD4002_SS = 1;
}


#define SPEECH_NUM 2        //分两段录音,ISD4002-120 为 120 s,每秒 5 段行地址,共 600
                            //段行地址
#define SPEECH_LEN 60       //每段 60 s
/* ---------------------------------------------
                    ISD4002 录音程序
---------------------------------------------- */
void speech_record(void)
{
    uchar i;
    ISD4002_powerup();
    delay1ms(25);
    ISD4002_powerup();
    delay1ms(25);
    for (i = 0;i<SPEECH_NUM;i++)
        {
        while (BUTTON1);
        REC_DISP_DATA[3] = DISP_TAB[i];              //录音段显示
        _write(0x90,15,REC_DISP_DATA);               //液晶第 2 行显示录音段位置
        ISD4002_record(i * (5 * SPEECH_LEN));        //发录音命令
        LED = 0;                                     //录音指示灯开
        delay1ms(SPEECH_LEN * 1000);                 //录音等待
        ISD4002_stop();                              //发录音停止命令
        LED = 1;                                     //录音指示灯关
        _write(0x90,15,REC_CON);                     //显示"请继续录音"
        }
    _write(0x90,15,REC_END);                         //显示"录音结束"
```

```c
}
/* ---------------------------------------------
                   ISD4002 放音程序
--------------------------------------------- */
void speech_play(void)
{
        uchar i;
        ISD4002_powerup();
        delay1ms(25);
        ISD4002_powerup();
        delay1ms(25);
        for (i = 0;i<SPEECH_NUM;i ++ )
            {
            while (BUTTON2);
            PLAY_DISP_DATA[3] = DISP_TAB[i];        //录音段显示
            _write(0x90,15,PLAY_DISP_DATA);         //液晶第 2 行显示录音段位置
            ISD4002_play(i * (5 * SPEECH_LEN));     //发放音命令
            LED1 = 0;                               //放音指示灯开
            delay1ms(SPEECH_LEN * 1000);            //放音等待
            ISD4002_stop();                         //发停止命令
            LED1 = 1;                               //放音指示灯关
            _write(0x90,15,PLAY_CON);               //显示"请继续放音"
            }
    _write(0x90,15,PLAY_END);                       //显示"放音结束"
}
//
/* ---------------------------------------------
                   void main(void)
--------------------------------------------- */
void main(void)
{
//以下为液晶初始化程序
setup();                                            //液晶上电初始化
BLACK = 0;                                          //开液晶背光灯
_write(0x80,15,LOG_DATA);                           //初始开机画面,写液晶第 1 行
delay1ms(5000);                                     //延时约 5 s
BLACK = 1;                                          //关背光灯
//以下为录音、放音程序
while (1)
    {
        if (!BUTTON1)                               //检测录音开关
        {
```

```
            speech_record();                        //录音程序
        }
        if (!BUTTON2)                               //检测放音开关
        {
            speech_play();                          //放音程序
        }
    }
}
```

/* --
                        end
-- */

第 9 章 超声波测距器的设计

9.1 系统功能

超声波测距器可应用于汽车倒车、建筑施工工地以及一些工业现场的位置监控,也可用于如液位、井深、管道长度、物体厚度等的测量。其测量范围为 0.10 ～ 4.00 m,测量精度为 1 cm。测量时与被测物体无直接接触,能够清晰、稳定地显示测量结果。

9.2 设计方案

我们知道,由于超声波指向性强,能量消耗缓慢,在介质中传播的距离较远,因而超声波经常用于距离的测量。利用超声波检测距离设计比较方便,计算处理也较简单,并且在测量精度方面也能达到日常使用的要求。

超声波发生器可以分为两大类:一类是用电气方式产生超声波;另一类是用机械方式产生超声波。电气方式包括压电型、电动型等,机械方式有加尔统笛、液哨和气流旋笛等,它们所产生的超声波的频率、功率和声波特性各不相同,因而用途也各不相同。目前,在近距离测量方面较为常用的是压电式超声波换能器。

根据设计要求并综合各方面因素,本例决定采用 AT89C52 单片机作为主控制器,用动态扫描法实现 LED 数字显示,用单片机的定时器产生超声波驱动信号。超声波测距器的系统设计框图如图 9.1 所示。

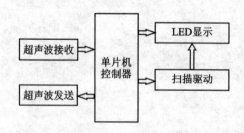

图 9.1 超声波测距器的系统设计框图

9.3 系统硬件电路的设计

硬件电路主要分为单片机系统及显示电路、超声波发射电路和超声波检测接收电路3部分。

9.3.1 单片机系统及显示电路

单片机采用AT89C52或其兼容系列。系统采用12 MHz高精度的晶振,以获得较稳定的时钟频率,并减小测量误差。单片机用P1.0引脚输出超声波换能器所需的40 kHz方波信号,利用外中断0口监测超声波接收电路输出的返回信号。显示电路采用简单实用的四位共阳LED数码管,段码用74LS244驱动,位码用PNP三极管9012驱动。单片机系统及显示电路如图9.2所示。

9.3.2 超声波发射电路

超声波发射电路原理图如图9.3所示。发射电路主要由反向器74LS04和超声波换能器构成,单片机P1.0引脚输出的40 kHz方波信号一路经一级反向器后送到超声波换能器的一个电极,另一路经两级反向器后送到超声波换能器的另一个电极,用这种推挽形式将方波信号加到超声波换能器两端可以提高超声波的发射强度。输出端采用两个反向器并联,用以提高驱动能力。上拉电阻R10、R11一方面可以提高反向器74LS04输出高电平的驱动能力;另一方面可以增加超声换能器的阻尼效果,以缩短其自由振荡的时间。

压电式超声波换能器是利用压电晶体的谐振来工作的。超声波换能器内部结构如图9.4所示,它有两个压电晶片和一个共振板。当它的两极外加脉冲信号,其频率等于压电晶片的固有振荡频率时,压电晶片将会发生共振,并带动共振板振动产生超声波,这时它就是一个超声波发生器;反之,如果两电极间未外加电压,当共振板接收到超声波时,将压迫压电晶片作振动,将机械能转换为电信号,这时它就成为超声波接收换能器了。超声波发射换能器与接收换能器在结构上稍有不同,使用时应分清器件上的标志。

9.3.3 超声波检测接收电路

集成电路CX20106A是一款红外线检波接收的专用芯片,常用于电视机红外遥控接收器。考虑到红外遥控常用的载波频率38 kHz与测距的超声波频率40 kHz较为接近,因此可以利用它制作超声波检测接收电路,如图9.5所示。实验证明,用CX20106A接收超声波(无信号时输出高电平)具有很高的灵敏度和较强的抗干扰能力。适当地更改电容C4的大小,可以改变接收电路的灵敏度和抗干扰能力。

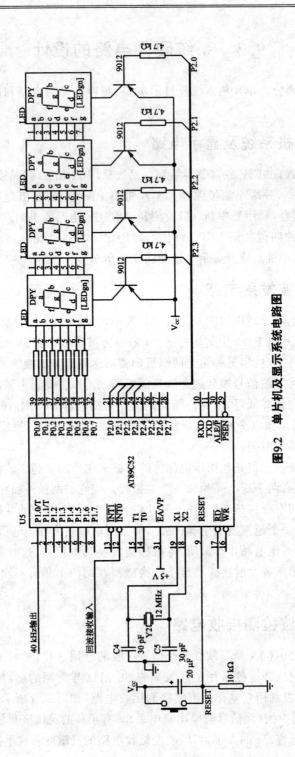

图9.2 单片机及显示系统电路图

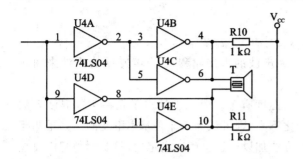
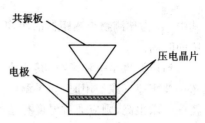

图 9.3　超声波发射电路原理图　　　　图 9.4　超声波换能器内部结构图

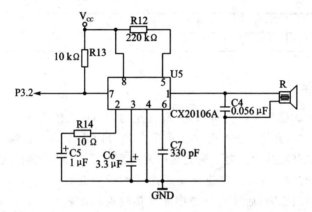

图 9.5　超声波检测接收电路图

9.4　系统程序的设计

超声波测距器的软件设计主要由主程序、超声波发生子程序、超声波接收中断程序及显示子程序组成。由于 C 语言程序有利于实现较复杂的算法,汇编语言程序则具有较高的效率并且容易精确计算程序运行的时间,而超声波测距器的程序既有较复杂的计算(计算距离时),又要求精确计算程序运行时间(超声波测距时),所以控制程序可采用 C 语言和汇编语言混合编程。下面对超声波测距器的算法、主程序、超声波发生子程序和超声波接收中断程序逐一介绍。

9.4.1　超声波测距器的算法设计

图 9.6 所示为超声波测距的原理,即超声波发生器 T 在某一时刻发出一个超声波信号,当这个超声波遇到被测物体后反射回来,就会被超声波接收器 R 接收到。这样,只要计算出从发出超声波信号到接收到返回信号所用的时间,就可算出超声波发生器与反射物体的距

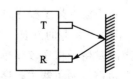

图 9.6　超声波测距原理图

离。该距离的计算公式如下：

$$d = s/2 = (v \times t)/2 \tag{9-1}$$

其中：d 为被测物与测距器的距离；s 为声波的来回路程；v 为声速；t 为声波来回所用的时间。

超声波也是一种声波，其声速 v 与温度有关。表 9.1 列出了几种不同温度下的超声波声速。在使用时，如果温度变化不大，则可认为声速是基本不变的。如果测距精度要求很高，则应通过温度补偿的方法加以校正。声速确定后，只要测得超声波往返的时间，即可求得距离。

表 9.1 不同温度下超声波声速表

温度/℃	−30	−20	−10	0	10	20	30	100
声速/(m·s^{-1})	313	319	325	323	338	344	349	386

9.4.2 主程序

主程序首先要对系统环境初始化，设置定时器 T0 工作模式为 16 位定时/计数器模式，置位总中断允许位 EA 并对显示端口 P0 和 P2 清 0；然后调用超声波发生子程序送出一个超声波脉冲。为了避免超声波从发射器直接传送到接收器引起的直射波触发，需要延时约 0.1 ms(这也就是超声波测距器会有一个最小可测距离的原因)后才打开外中断 0 接收返回的超声波信号。由于采用的是 12 MHz 的晶振，计数器每计一个数就是 1 μs，所以当主程序检测到接收成功的标志位后，将计数器 T0 中的数(即超声波来回所用的时间)按式(9-2)计算，即可得被测物体与测距器之间的距离。设计时取 20 ℃ 时的声速为 344 m/s，则有

$$d = (v \times t)/2 = (172 T_0 /10000) \text{ cm} \tag{9-2}$$

其中：T_0 为计数器 T0 的计数值。

测出距离后，结果将以十进制 BCD 码方式送往 LED 显示约 0.5 s，然后再发超声波脉冲重复测量过程。图 9.7 所示为主程序流程图。

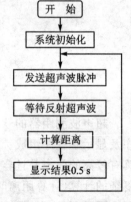

图 9.7 主程序流程图

9.4.3 超声波发生子程序和超声波接收中断程序

超声波发生子程序的作用是通过 P1.0 引脚发送两个左右的超声波脉冲信号(频率约 40 kHz 的方波)，脉冲宽度为 12 μs 左右，同时把计数器 T0 打开进行计时。超声波发生子程序较简单，但要求程序运行时间准确，所以采用汇编语言编程。

超声波测距器主程序利用外中断 0 检测返回超声波信号，一旦接收到返回超声波信号(即 INT0 引脚出现低电平)，立即进入超声波接收中断程序。进入该中断后，

就立即关闭计时器 T0,停止计时,并将测距成功标志字赋值 1。

如果当计时器溢出时还未检测到超声波返回信号,则定时器 T0 溢出中断将外中断 0 关闭,并将测距成功标志字赋值 2,以表示本次测距不成功。

9.5 调试及性能分析

9.5.1 调 试

超声波测距器的制作和调试都较为简单,其中超声波发射和接收采用 $\phi15$ 的超声波换能器 TCT40-10F1(T 发射)和 TCT40-10S1(R 接收),中心频率为 40 kHz,安装时应保持两换能器中心轴线平行并相距 4~8 cm,其余元件无特殊要求。若能将超声接收电路用金属壳屏蔽起来,则可提高抗干扰性能。根据测量范围要求不同,可适当地调整与接收换能器并接的滤波电容 C4 的大小,以获得合适的接收灵敏度和抗干扰能力。

硬件电路制作完成并调整好后,便可将程序编译好下载到单片机试运行。根据实际情况,可以修改超声波发生子程序每次发送的脉冲个数和两次测量的间隔时间,以适应不同距离的测量需要。

9.5.2 性能指标

根据文中电路参数和程序,测距器可测量的范围为 0.07~5.50 m。实验中,对测量范围为 0.07~2.50 m 的平面物体做了多次测试,测距器的最大误差不超过 1 cm,重复性很好。

9.6 源程序清单

9.6.1 汇编源程序清单

```
;************************************
;*              超声波测距器              *
;*         采用 AT89C52  12 MHz 晶振      *
;*            采用共阳 LED 显示器           *
;************************************
;测距范围为 0.07~4 m,堆栈在 4FH 以上,20H 用于标志
;显示缓冲单元在 40H~43H,内存 44H~46H 用于计算距离

     VOUT      EQU    P1.0              ;脉冲输出端口

;**************** 中断入口程序 ****************
```

```
            ORG     0000H
            LJMP    START
            ORG     0003H
            LJMP    PINT0
            ORG     000BH
            LJMP    INTT0
            ORG     0013H
            RETI
            ORG     001BH
            LJMP    INTT1
            ORG     0023H
            RETI
            ORG     002BH
            RETI
;******************* 主程序 *******************
START:      MOV     SP,#4FH
            MOV     R0,#40H         ;40H~43H 为显示数据存放单元(40H 为最高位)
            MOV     R7,#0BH
CLEARDISP:  MOV     @R0,#00H
            INC     R0
            DJNZ    R7,CLEARDISP
            MOV     20H,#00H
            MOV     TMOD,#21H       ;T1 为 8 位自动重装模式,T0 为 16 位定时器
            MOV     TH0,#00H        ;65 ms 初值
            MOV     TL0,#00H
            MOV     TH1,#0F2H       ;40 kHz 初值
            MOV     TL1,#0F2H
            MOV     P0,#0FFH
            MOV     P1,#0FFH
            MOV     P2,#0FFH
            MOV     P3,#0FFH
            MOV     R4,#04H         ;超声波脉冲个数控制(为赋值的一半)
            SETB    PX0
            SETB    ET0
            SETB    EA
            SETB    TR0             ;开启测距定时器
START1:     LCALL   DISPLAY
            JNB     00H,START1      ;收到反射信号时标志位为 1
            CLR     EA
            LCALL   WORK            ;计算距离子程序
            SETB    EA
```

第 9 章 超声波测距器的设计

```
               CLR    00H
               SETB   TR0                ;重新开启测距定时器
               MOV    R2,#64H            ;测量间隔控制(约 4 ms×100 = 400 ms)
LOOP:          LCALL  DISPLAY
               DJNZ   R2,LOOP
               SJMP   START1
```

;******************** 中断程序 ********************
;T0 中断,65 ms 中断一次

```
INTT0:         CLR    EA
               CLR    TR0
               MOV    TH0,#00H
               MOV    TL0,#00H
               SETB   ET1
               SETB   EA
               SETB   TR0                ;启动计数器 T0,用以计算超声波来回时间
               SETB   TR1                ;开启发超声波用定时器 T1
OUT:           RETI
```

;T1 中断,发超声波用

```
INTT1:         CPL    VOUT
               DJNZ   R4,RETIOUT
               CLR    TR1                ;超声波发送完毕,关 T1
               CLR    ET1
               MOV    R4,#04H
               SETB   EX0                ;开启接收回波中断
RETIOUT:       RETI
```

;外中断 0,收到回波时进入

```
PINT0:         CLR    TR0                ;关计数器
               CLR    TR1
               CLR    ET1
               CLR    EA
               CLR    EX0
               MOV    44H,TL0            ;将计数值移入处理单元
               MOV    45H,TH0
               SETB   00H                ;接收成功标志
               RETI
```

;******************** 显示程序 ********************
;40H 为最高位,43H 为最低位,先扫描高位

```
DISPLAY:       MOV    R1,#40H;G
               MOV    R5,#0F7H;G
PLAY:          MOV    A,R5
               MOV    P0,#0FFH
```

```
            MOV     P2,A
            MOV     A,@R1
            MOV     DPTR,#TAB
            MOVC    A,@A+DPTR
            MOV     P0,A
            LCALL   DL1MS
            INC     R1
            MOV     A,R5
            JNB     ACC.0,ENDOUT;G
            RR      A
            MOV     R5,A
            AJMP    PLAY
ENDOUT:     MOV     P2,#0FFH
            MOV     P0,#0FFH
            RET
TAB:        DB      0C0H,0F9H,0A4H,0B0H,99H,92H,82H,0F8H,80H,90H,0FFH,88H,0BFH
;共阳段码表          "0","1","2","3","4","5","6","7","8","9","不亮","A","-"
;********************延时程序********************
DL1MS:      MOV     R6,#14H
DL1:        MOV     R7,#19H
DL2:        DJNZ    R7,DL2
            DJNZ    R6,DL1
            RET
;*********距离计算程序（距离=计数值×17/1000 cm）*********
work:       PUSH    ACC
            PUSH    PSW
            PUSH    B
            MOV     PSW,#18h
            MOV     R3,45H
            MOV     R2,44H
            MOV     R1,#00D
            MOV     R0,#17D
            LCALL   MUL2BY2
            MOV     R3,#03H
            MOV     R2,#0E8H
            LCALL   DIV4BY2
            LCALL   DIV4BY2
            MOV     40H,R4
            MOV     A,40H
            JNZ     JJ0
            MOV     40H,#0AH        ;最高位为0,不点亮
```

```
JJ0:        MOV     A,R0
            MOV     R4,A
            MOV     A,R1
            MOV     R5,A
            MOV     R3,#00D
            MOV     R2,#100D
            LCALL   DIV4BY2
            MOV     41H,R4
            MOV     A,41H
            JNZ     JJ1
            MOV     A,40H                       ;次高位为0,先看最高位是否为不亮
            SUBB    A,#0AH
            JNZ     JJ1
            MOV     41H,#0AH                    ;最高位不亮,次高位也不亮
JJ1:        MOV     A,R0
            MOV     R4,A
            MOV     A,R1
            MOV     R5,A
            MOV     R3,#00D
            MOV     R2,#10D
            LCALL   DIV4BY2
            MOV     42H,R4
            MOV     A,42H
            JNZ     JJ2
            MOV     A,41H                       ;次次高位为0,先看次高位是否为不亮
            SUBB    A,#0AH
            JNZ     JJ2
            MOV     42H,#0AH                    ;次高位不亮,次次高位也不亮
JJ2:        MOV     43H,R0
            POP     B
            POP     PSW
            POP     ACC
            RET
;***************2字节无符号数乘法程序 ***************
;R7R6R5R4 ≤ R3R2 × R1R0
MUL2BY2:    CLR     A
            MOV     R7,A
            MOV     R6,A
            MOV     R5,A
            MOV     R4,A
            MOV     46H,#10H
```

```
MULLOOP1:   CLR     C
            MOV     A,R4
            RLC     A
            MOV     R4,A
            MOV     A,R5
            RLC     A
            MOV     R5,A
            MOV     A,R6
            RLC     A
            MOV     R6,A
            MOV     A,R7
            RLC     A
            MOV     R7,A
            MOV     A,R0
            RLC     A
            MOV     R0,A
            MOV     A,R1
            RLC     A
            MOV     R1,A
            JNC     MULLOOP2
            MOV     A,R4
            ADD     A,R2
            MOV     R4,A
            MOV     A,R5
            ADDC    A,R3
            MOV     R5,A
            MOV     A,R6
            ADDC    A,#00H
            MOV     R6,A
            MOV     A,R7
            ADDC    A,#00H
            MOV     R7,A
MULLOOP2:   DJNZ    46H,MULLOOP1
            RET

;****************4字节/2字节无符号数除法程序****************
;R7R6R5R4/R3R2 = R7R6R5R4(商)…R1R0(余数)
DIV4BY2:    MOV     46H,#20H
            MOV     R0,#00H
            MOV     R1,#00H
DIVLOOP1:   MOV     A,R4
            RLC     A
```

第 9 章 超声波测距器的设计

```
              MOV     R4,A
              MOV     A,R5
              RLC     A
              MOV     R5,A
              MOV     A,R6
              RLC     A
              MOV     R6,A
              MOV     A,R7
              RLC     A
              MOV     R7,A
              MOV     A,R0
              RLC     A
              MOV     R0,A
              MOV     A,R1
              RLC     A
              MOV     R1,A
              CLR     C
              MOV     A,R0
              SUBB    A,R2
              MOV     B,A
              MOV     A,R1
              SUBB    A,R3
              JC      DIVLOOP2
              MOV     R0,B
              MOV     R1,A
DIVLOOP2:     CPL     C
              DJNZ    46H,DIVLOOP1
              MOV     A,R4
              RLC     A
              MOV     R4,A
              MOV     A,R5
              RLC     A
              MOV     R5,A
              MOV     A,R6
              RLC     A
              MOV     R6,A
              MOV     A,R7
              RLC     A
              MOV     R7,A
              RET
              END                      ;程序结束
```

9.6.2　C 源程序清单

以下是采用汇编和 C52 混合编程的源程序：

```c
;/* 文件 1:cscjmain.c */
/* - - - - - - - - - - - - - - - - - - - -
超声波测距器单片机程序(MCU AT89C52 XAL 12 MHz)
- - - - - - - - - - - - - - - - - - - - */
#include <reg52.h>
#define uchar unsigned char
#define uint unsigned int
#define ulong unsigned long
extern void cs_t(void);
extern void delay(uint);
extern void display(uchar *);
data uchar testok;
/* 主程序 */
void main(void)
{
    data uchar dispram[5];
    data uint i;
    data ulong time;
    P0 = 0xFF;
    P2 = 0xFF;
    TMOD = 0x11;
    IE = 0x80;
    while (1)
    {
        cs_t();
        delay(1);
        testok = 0;
        EX0 = 1;
        ET0 = 1;
        while (! testok) display(dispram);
        if (1 == testok)
        {
            time = TH0;
            time = (time<<8) | TL0;
            time *= 172;
            time /= 10000;
            dispram[0] = (uchar) (time % 10);
            time /= 10;
            dispram[1] = (uchar) (time % 10);
```

```c
        time/= 10;
        dispram[2] = (uchar)(time % 10);
        dispram[3] = (uchar)(time/10);
        if (0 == dispram[3]) dispram[3] = 17;
    } else
    {
        dispram[0] = 16;
        dispram[1] = 16;
        dispram[2] = 16;
        dispram[3] = 16;
    }
    for (i = 0; ik<300; i++) display(dispram);
    }
}
/* 超声波接收程序(外中断 0) */
void cs_r(void) interrupt 0
{
    TR0 = 0;
    ET0 = 0;
    EX0 = 0;
    testok = 1;
}
/* 超时清除程序(内中断 T0) */
void overtime(void) interrupt 1
{
    EX0 = 0;
    TR0 = 0;
    ET0 = 0;
    testok = 2;
}
```

;/* 文件 2:cs_t.asm */
;--------超声波发生子程序(12 MHz 晶振,38.5 Hz)--------

```
                        NAME        CS_T
? PR? CS_T? CS_T        SEGMENT     CODE
                        PUBLIC      CS_T
                        RSEG        ? PR? CS_T? CS_T
CS_T:                   PUSH        ACC
                        MOV         TH0,#00H
                        MOV         TL0,#00H
                        MOV         A,#4D
                        SETB        TR0
CS_T1:                  CPL         P1.0
```

```
                    NOP
                    NOP
                    NOP
                    NOP
                    NOP
                    NOP
                    NOP
                    NOP
                    NOP
                    DJNZ        ACC,CS_T1
                    POP         ACC
                    RET
                    END
```

;/*文件3:display.asm*/
;----------四位共阳LED动态扫描显示程序----------
;P0为段码口,P2为位选口(低电平有效)
;参数为要显示的字符串指针

```
                    NAME        DISPLAY
? PR? _DISPLAY? DISPLAY  SEGMENT     CODE
? CO? _DISPLAY? DISPLAY  SEGMENT     DATA
                    EXTRN       CODE(_DELAY)
                    PUBLIC      _DISPLAY
                    RSEG        ? CO? _DISPLAY? DISPLAY
? _DISPLAY? BYTE:
DISPBIT:            DS          1
DISPNUM:            DS          1
                    RSEG        ? PR? _DISPLAY? DISPLAY
_DISPLAY:           PUSH        ACC
                    PUSH        DPH
                    PUSH        DPL
                    PUSH        PSW
                    INC         DISPNUM
                    MOV         A,DISPNUM
                    CJNE        A,#4D,DISP1
DISP1:              JC          DISP2
                    MOV         DISPNUM,#00H
                    MOV         DISPBIT,#0FEH
DISP2:              MOV         A,R1
                    ADD         A,DISPNUM
                    MOV         R0,A
```

第 9 章 超声波测距器的设计

```
            MOV     A,@R0
            MOV     DPTR,#DISPTABLE
            MOVC    A,@A+DPTR
            MOV     P0,A
            MOV     A,DISPNUM
            CJNE    A,#2D,DISP3
            CLR     P0.7
DISP3:      MOV     P2,DISPBIT
            MOV     R6,#00H
            MOV     R7,#0AH
            LCALL   _DELAY
            MOV     P0,#0FFH
            MOV     P2,#0FFH
            MOV     A,DISPBIT
            RL      A
            MOV     DISPBIT,A
            POP     PSW
            POP     DPL
            POP     DPH
            POP     ACC
            RET
DISPTABLE:  DB      0C0H,0F9H,0A4H,0B0H,99H,92H,82H,0F8H,80H,90H,
                    88H,83H,0C6H,0A1H,86H,8EH,0BFH,0FFH
;                   "0","1","2","3","4","5","6","7","8","9","A",
                    "B","C","D","E","F","-",""
            END
```

;/* 文件 4:delay.asm */
;————————延时 100 个机器周期×参数(1~65535)————————

```
                    NAME    DELAY
?PR?_DELAY?DELAY    SEGMENT CODE
                    PUBLIC  _DELAY
                    RSEG    ?PR?_DELAY?DELAY
_DELAY:             PUSH    ACC             ;2
                    MOV     A,R7            ;1
                    JZ      DELA1           ;2
                    INC     R6              ;1
DELA1:              MOV     R5,#50D         ;2
                    DJNZ    R5,$            ;2
                    DJNZ    R7,DELA1        ;2
                    DJNZ    R6,DELA1        ;2
                    POP     ACC             ;2
                    RET                     ;2
                    END
```

第 10 章 简易 LCD 示波器的设计

10.1 系统功能

简易 LCD 示波器要求用液晶显示器显示 30~70 Hz 范围内的工频 220 V 交流输入电压的信号波形,并可以调整波形的幅度及水平扫描频率。

10.2 设计方案

数字示波器是实验室里最常用的信号测试设备,它可以直观地观察信号的波形并能测量信号的幅度、周期、频率等参数。考虑到作为单片机课程设计内容的教学时数及难度,简易 LCD 示波器课程设计重点要求掌握波形的模/数转换和 LCD 液晶显示器波形显示的方法。为此,设计中采用了带内部高速 A/D 转换器的单片机 STC12C5A16S2 作为控制器,LCD 液晶显示器采用了 128×64 点阵的 JXD12864F-1,输入端采用电压变换器及电平转换电路进行简单的处理。

简易 LCD 示波器系统设计框图如图 10.1 所示。

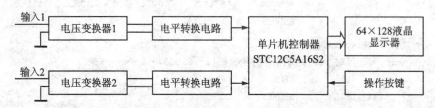

图 10.1 简易 LCD 示波器系统设计框图

10.3 系统硬件电路的设计

简易 LCD 示波器电路原理图如图 10.2 所示。两路交流电压信号分别经电压变换器进行电平变换,成为 0~5 V 的正极性交流信号电压。单片机的 A/D 输入端口 P1.0、P1.1 通过程序控制选择一路或两路进行 A/D 转换,将实时采集的电压数字量存放在内部的 RAM 采集表中;当采集完成后去刷新显示缓冲器数据表,处理完表中所有数据的序号、数值大小与对应的 LCD 显示屏中的 X 方向(水平时间轴)、Y 方向(电压幅度)的点坐标的位置后,将波形信息数据写入 LCD,完成电压信号波形的显

第 10 章 简易 LCD 示波器的设计

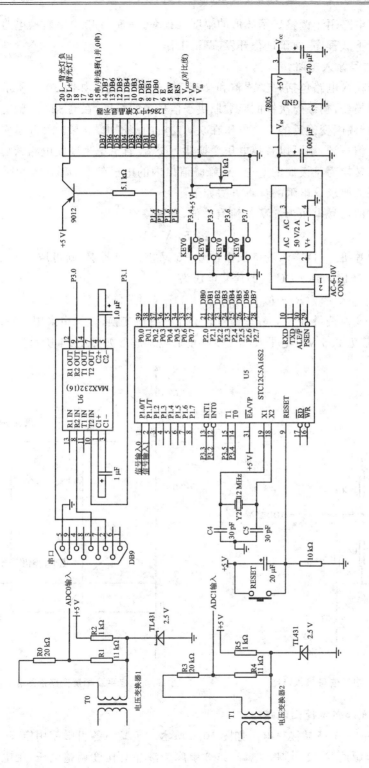

图 10.2 简易 LCD 示波器电路原理图

示过程。按键电路用于调整显示波形的幅度及频率。RS-232电平转换电路用于计算机串口直接下载程序,便于进行开发、调试工作。

1. 电压信号输入电路

电压信号输入电路包括幅度调整和交流/直流变换,电路如图10.3所示。幅度调整用电压变换器,初级与次级的匝数取决于信号源的幅度,在测220 V的交流电压时,要求在进行幅度变换后的交流电压在5 V左右的范围内,如果信号源是实验室里的小幅度电压信号源,则可以省掉电压变换器。交流/直流电平变换电路的功能是将输入交流信号转换为正极性的0~5 V峰峰值以内的电压信号,以满足单片机A/D输入端口电平的要求。电阻的计算方法如下:

设输入端的信号电压为5 V交流有效值,则其峰峰值为

$$V_{P-P} = (2 \times 1.414 \times 5) \text{ V} = 14.14 \text{ V}$$

要满足单片机A/D输入端口正极性0~5 V电压的要求,则可取 $R_3 = 20 \text{ k}\Omega$, $R_4 = 11 \text{ k}\Omega$,这样输入到ADC口的信号电压为

$$V_{P-P} \times R_4/(R_3 + R_4) = [14.14 \times 11 \div (20+11)] \text{ V} = 5.02 \text{ V}$$

直流电平变换按最大5 V峰峰值进行交流—直流变换,设计中采用TL431三端稳压管,稳压值为2.5 V。电平及幅度变换前后输入波形如图10.4所示。

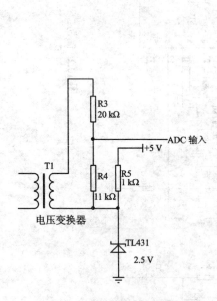

图10.3 电压信号输入电路

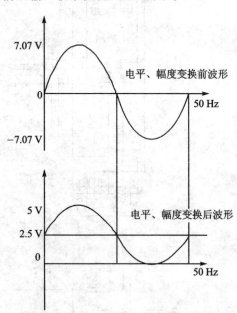

图10.4 电平及幅度变换前后输入波形

2. LCD液晶显示接口电路

JXD12864F-1液晶接口电路如图10.5所示。接口中各引脚使用定义分别为:第1引脚为电源地;第2引脚为电源;第3引脚为液晶对比度调整电压,用电位器控

制液晶屏的对比度,一般电压在 3~4.5 V;第 4 引脚为数据/命令控制;第 5 引脚为读/写控制;第 6 引脚为使能口(片选);第 7~14 引脚为并行数据传送口;第 15 引脚为串行/并行通信选择口;第 16~18 引脚为空脚;第 19、20 引脚分别为背光灯的电源引脚,采用三极管 9012 进行驱动。为了提高 LCD 示波器的显示速度,单片机与液晶显示器采用并行连接方式,通过第 15 引脚电平的选择决定接口是并行通信还是串行通信。在模块使用时要注意与第 15 引脚相连的焊点的选择,将与电源相对的焊点焊上,使第 15 引脚接 5 V 电压。

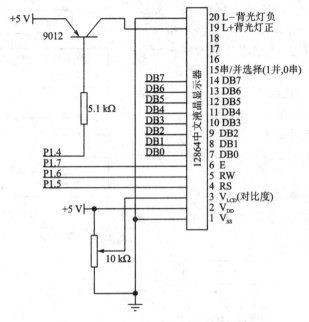

图 10.5　JXD12864F-1 液晶接口电路

10.4　系统程序的设计

1. LCD 液晶显示器实现画图功能的方法

JXD12864F-1 液晶显示器具有 4 行中西文或点阵显示功能,文本显示功能与前面实例中使用的 12232F 液晶显示器相同,这里重点介绍显示波形时的画图功能。点阵画图指令属于扩充指令集,要使用这些指令必须在液晶初始化之后写命令字(0x34)进入扩充指令集设定状态。整个画图(画波形)的过程按以下步骤进行:

① 初始化 LCD(一般上电时做初始化);
② 写入指令(0x34),进入扩充指令集设定状态;
③ 设定 GDRAM 的行列地址,然后连续写入图形数据;
④ 写入指令 0x36,打开绘图开关,使其正常显示图形;
⑤ 写入指令 0x30,返回基本指令集设定状态。

JXD12864F-1 液晶显示器将屏幕分成上、下两部分,每部分的垂直坐标为 00H～1FH(两行共 32 个点),水平地址为 00H～0FH。在向液晶 GDRAM 中写入要显示的波形数据时,首先要指定 X(水平方向)及 Y(垂直方向)坐标,然后写入数据。在写入数据时,LCD 会先写高位字节,再写低位字节(每个字节 8 个点);然后 LCD 会自动把水平坐标定位到同一行下一个地址,接着写入波形数据,依次类推,当写满 16 次后,第 00 行就全写满了;接着在程序中要人为地把垂直方向坐标(Y 轴)的地址加 1,然后重复上述过程,直到全部数据写完为止。

在写入定位初始 X、Y 坐标时,要首先设垂直地址,接着设定水平地址,这两个指令是连续写入 LCD 的。其方法是进入扩充指令集设定状态后,将 RS 引脚置低电平,RW 引脚置低电平,接着连续写入上面两个坐标字节(垂直地址在前,列地址在后)。

2. 主程序流程图

简易 LCD 示波器主程序流程图如图 10.6 所示。

3. 画波形程序流程图

画波形程序的主要流程图如图 10.7 所示。

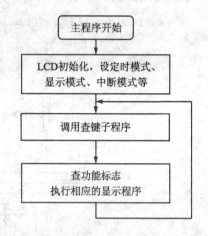

图 10.6 简易 LCD 示波器主程序流程图

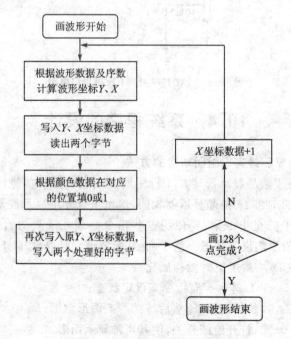

图 10.7 画波形程序的主要流程图

4. 波形数据采集中断程序流程图

波形数据采集中断程序流程图如图 10.8 所示。

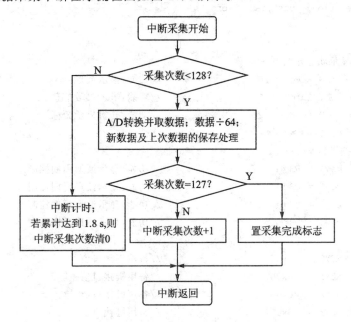

图 10.8　波形数据采集中断程序流程图

10.5　调试及性能分析

调试时,先用低频信号源输出正弦波信号,幅度一般控制在峰峰值 5 V 以内,然后观察液晶屏波形的显示情况,待波形显示正常后,再调试幅度控制程序、周期控制程序等子程序,最后在确认 220 V 电压变换器输出在规定范围值内后,连接工频 220 V 交流电压,连接时要特别注意人身安全。经调试后,简易 LCD 示波器能进行正常的 220 V 交流输入电压信号波形的显示,并可以调整波形的幅度及水平扫描频率。

10.6　源程序清单

```
//***************************************************//
//          12864 LCD 示波器演示程序
//          STC12C5A16S2      12 MHz 晶振
//***************************************************//
//头文件
# include "STC12C5A16S2.H"
# include"intrins.h"
# include"math.h"
```

```c
#define uchar unsigned char
#define uint  unsigned int
#define ulong unsigned long
//
//========================
//****显示状态定义****
#define STA_Wave   0          //功能:显示波形
#define STA_Info   1          //功能:显示参数信息
#define STA_Logo   2          //功能:开机显示徽标
//========================
//****按键定义****
sbit    key0 = P3^4;          //功能:显示波形周期控制
sbit    key1 = P3^5;          //功能:显示波形周期控制
sbit    key2 = P3^6;          //功能:显示波形幅度控制
sbit    key3 = P3^7;          //功能:显示波形幅度控制
//
#define KEY0    0x0E          //备用键接口值
#define KEY1    0x0D          //备用键接口值
#define KEY2    0x0B          //备用键接口值
#define KEY3    0x07          //备用键接口值
//
#define KEY0_S   0xE0         //KEY0 键值
#define KEY1_S   0xD0         //KEY1 键值
#define KEY2_S   0xB0         //KEY2 键值
#define KEY3_S   0x70         //KEY3 键值
#define KEY_NONE  0           //无按键
#define lcddata    P2         //12864 液晶数据口
sbit busy = lcddata^7;        //12864 液晶"忙"接口
//
uchar STATE = 0;              //功能判断(数据为 0~2)
uchar MSG_KEY = KEY_NONE;     //键值存放
uchar key_times = 0;          //键扫描时间
uchar key_keep_times = 0;     //按键按住时间
uchar key_done = 0;           //按键事件有无做好标志
uchar read_key_times = 0;     //扫描键盘时间
uchar draw_rigol = 0;         //画时间
//
uchar data X_TIMES = 0;       //X 坐标
uchar data T_val = 12;        //波形采样周期控制初值
uchar data T_min = 10;        //波形采样周期控制最小值
uchar data T_max = 150;       //波形采样周期控制最大值
uchar data Margin = 16;       //显示波形幅度控制(范围为 1~16)
```

```
uint data ADC_RES_DATA = 0;              //10 位 A/D 转换数据存放
uint data con05s = 0;                    //0.5 s 计数器
long int data  bo_delay = 0;             //波形采集计时
uchar xdata   LAST_Y[128];               //上次采集数据存放
uchar xdata   NOW_Y[128];                //即时采集数据存放
bit data   draw_bo_is_ok = 0;            //画波形完成标志
//
//12864 液晶显示控制口
sbit light = P1^4;                       //背光灯控制口
sbit rs = P1^5;                          //1 表示数据,0 表示命令
sbit rw = P1^6;                          //1 表示读,0 表示写
sbit e = P1^7;                           //使能口,下降沿有效
//信息表
uchar   code    disdata1[15] = {"单片机小示波器 "};//开机显示画面
uchar   code    disdata2[15] = {"浙江海洋学院     "};//开机显示画面
//函数定义
void write_data(uchar dat_d);
void write_com(uchar dat_d);
void clrram_lcd (void);
void init_lcd(void);
void busy_lcd(void);
void rev_row_lcd(uchar row);
void rev_co_lcd(uchar row,uchar col,uchar mode);
void clr_lcd(void);
void wr_co_lcd(uchar row,uchar col,uchar lcddata1,uchar lcddtta2);
void wr_row_lcd(uchar row,char * p);
//
//以下为程序区
//
// ************1/12 ms 延时程序 *************
delay1ms(uint t)
{
int i,j;
for(i = 0;i<t;i++ )
    for(j = 0;j<64;j++ )
    ;
}
//
// **********液晶检测忙状态程序 ***************
//在写入之前必须执行
void busy_lcd(void)
{
```

```c
    lcddata = 0xFF;
    rs = 0;
    rw = 1;
    e = 1;
    while(busy == 1);
    e = 0;
}
//
/**********向 12232F 写 1 字节程序 ************/
void write_data(uchar dat_d)
{
//并口使用时
    busy_lcd();
    rs = 1;
    rw = 0;
    lcddata = dat_d;
    e = 1;
    ;
    e = 0;
}
//
/*********向 12232F 写 1 个命令字节程序 ***********/
void write_com(uchar com_d)
{
 //并口使用时
        busy_lcd();
        rs = 0;
        rw = 0;
        lcddata = com_d;
        e = 1;
        ;
        e = 0;
}
//
// **********12864 LCD 初始化程序 **************
void  setup()
{
        write_com(0x0C);                /*开显示,关游标*/
        clrram_lcd();
        delay1ms(480);                  //延时 40 ms 以上*/
}
//
```

```c
// **********填充液晶 GDRAM 全为空格程序 **********
void clrram_lcd (void)
{
    write_com(0x30);
    write_com(0x01);
}
//
// **********液晶写入函数 **********
void _write(uchar _wz,uchar _len,uchar * _adr)
{
    uchar m;
    //q = _wz;
    write_com(_wz);                        //写入字定位(起始位)
    for (m = 0; m < _len; m ++ )
    {write_data( * ( _adr + m));}
}
//
// ************指定要显示字符的坐标程序 **************
void gotoxy(unsigned char y, unsigned char x)
{
    if(y == 1)
        write_com(0x80|x);                 //wr_i_lcd(0x80|x);
    if(y == 2)
        write_com(0x90|x);
    if(y == 3)
        write_com((0x80|x) + 8);
    if(y == 4)
        write_com((0x90|x) + 8);
}
//
// ************液晶显示字符串程序 ***************
void print(uchar * str)
{
    while( * str! = '\0')
    {
        //wr_d_lcd( * str);
        write_data( * str);
        str ++ ;
    }
}
//
// **************延时子程序 ********************
```

```c
void delay(unsigned int t)
{
    unsigned int i,j ;
    for(i = 0;i<t;i ++ )
    for(j = 0;j<10;j ++ );
}
//
// ***************清整个 GDRAM 空间程序 ***************
void clrgdram()
{
    unsigned char x,y ;
    for(y = 0;y<64;y ++ )
    for(x = 0;x<16;x ++ )
    {
        write_com(0x34);
        write_com(y + 0x80);            //行地址(Y 坐标)
        write_com(x + 0x80);            //列地址(X 坐标)
        write_com(0x30);
        write_data(0x00);
        write_data(0x00);
    }
}
//
// ***************清屏幕程序 ******************
void clrscreen()
{
    write_com(0x01);
    delay(10);
}
//
// ***************读 12864 字节程序 ******************
unsigned char ReadByte(void)
{
    unsigned char byReturnValue ;
    busy_lcd();
    lcddata = 0xFF;
    rs = 1 ;
    rw = 1 ;
    e = 1 ;
    e = 1 ;
    byReturnValue = lcddata;
    e = 0 ;
```

```c
        return byReturnValue ;
}
//
// ****************画点(波形)子程序 *********************
void DrawPoint(unsigned char X,unsigned char Y,unsigned char Color)
{
    unsigned char Row,Tier,Tier_bit ;
    unsigned char ReadOldH,ReadOldL ;
    write_com(0x34);                        //进入扩展指令
    write_com(0x36);                        //打开绘图开关
    Tier = X>>4 ;
    Tier_bit = X&0x0F;
    if(Y<32)
    {
        Row = Y ;                           //在上两行屏的 Y 坐标
    }
    else
    {
        Row = Y - 32 ;                      //在下两行屏的 Y 坐标
        Tier += 8 ;                         //在下两行屏的 X 坐标修正(1/2
                                            //行为 80//90,3/4 行为 88/98)
    }
    write_com(Row + 0x80);                  //写入 Y 坐标数据
    write_com(Tier + 0x80);                 //写入 X 坐标数据
    ReadByte();
    ReadOldH = ReadByte();                  //读入 GDRAM 数据
    ReadOldL = ReadByte();                  //读入 GDRAM 数据
    write_com(Row + 0x80);                  //重新设定坐标 Y
    write_com(Tier + 0x80);                 //重新设定坐标 X
    if(Tier_bit<8)
    {
        switch(Color)                       //左(高字节)8个点处理
        {
            case 0 :
            ReadOldH& = (~(0x01<<(7 - Tier_bit)));  //填 0
            break ;
            case 1 :
            ReadOldH| = (0x01<<(7 - Tier_bit));     //填 1
            break ;
            case 2 :
            ReadOldH^ = (0x01<<(7 - Tier_bit));
            break ;
```

```c
                default :
                break ;
            }
            write_data(ReadOldH);                    //写回 GDRAM
            write_data(ReadOldL);                    //写回 GDRAM
        }
        else
        {
            switch(Color)                            //右(低字节)8 个点处理
            {
                case 0 :
                ReadOldL& = (~(0x01<<(15 - Tier_bit))); //填 0
                break ;
                case 1 :
                ReadOldL| = (0x01<<(15 - Tier_bit));    //填 1
                break ;
                case 2 :
                ReadOldL^ = (0x01<<(15 - Tier_bit));
                break ;
                default :
                break ;
            }
            write_data(ReadOldH);                    //写回 GDRAM
            write_data(ReadOldL);                    //写回 GDRAM
        }
        write_com(0x30);
}
//
// ***************图片显示程序 *******************************
void v_Lcd12864DrawPicture_f( unsigned char code * pPicture )
{
    unsigned char i, j, k ;
    write_com( 0x34 ) ;
    write_com( 0x36 ) ;
    for( i = 0 ; i<2 ; i++ )
    {
        for( j = 0 ; j<32 ; j++ )
        {
            write_com( 0x80 + j ) ;
            if( i == 0 )
            {
                write_com( 0x80 ) ;
```

```c
            }
            else
            {
                write_com( 0x88 ) ;
            }
            for( k = 0 ; k<16 ; k++ )
            {
                write_data( *pPicture++ ) ;
            }
        }
    }
    write_com( 0x30 ) ;
}
//
/ *************STC12C5A16S2 ADC 电源开程序 ****************/
ADC_POW_ON()
{   ADC_CONTR = ADC_CONTR|0xE0;             //电源开,ADC 最快模式
    delay1ms(12);                           //延时 1 ms 以上
}
//
/ *************STC12C5A16S2 ADC 转换程序 ****************/
ADC_CHANG()                                 //A/D 转换
{
    //delay1ms(1);                          //延时 90 μs(通道转换后需稳定)
    _nop_();_nop_();                        //5 μs
    ADC_RES = 0x00;ADC_RESL = 0x00;ADC_CONTR = ADC_CONTR|0x08;//启动 ADC
    while((ADC_CONTR&0X10) == 0);           //ADC_CONTR.4 = 0 时,等待
    ADC_CONTR = ADC_CONTR&0xE7;             //清开始位及完成标志位
    ADC_RES_DATA = (ADC_RES * 4 + ADC_RESL);
}
//
// ************* 信息显示处理程序 ************
void Show_INFO()
{
}
//
// ************* 波形显示处理程序 ************
void draw_bo(){
    //
    uchar i;
    if(draw_bo_is_ok){
        for(i = 0;i<128;i++){
```

```c
                DrawPoint( i, NOW_Y[i], 1 );
                DrawPoint( i, LAST_Y[i], 0 );
            }
        draw_bo_is_ok = 0;
        }
}
//
// ****************按键处理程序************************
uchar ReadKey()
{
    if(!key_done)
    {
      switch(MSG_KEY)
        {
        case KEY0_S:
                    if(T_val > = T_max){T_val = T_max;}else{T_val = T_val + 10;}
                                                //显示波形周期控制
                    break;
        case KEY1_S:
                    if(T_val < = T_min){T_val = T_min;}else{T_val = T_val - 10;}
                                                //显示波形周期控制
                    break;
        case KEY2_S:
                    if(Margin > = 16){Margin = 16;}else{Margin ++ ;}
                                                //显示波形幅度变小
                    break;
        case KEY3_S:
                    if(Margin < = 8){Margin = 8;}else{Margin -- ;}
                                                //显示波形幅度变大
                    break;
        default:    break;
            }
        key_done = 1;                           //按键事件处理完成
    }
}
//
// ****************主程序***********************//
void   main()
{
    e = 0;                                      //12864 关闭
    light = 0;                                  //开液晶背光灯
    setup();                                    //液晶上电初始化
```

第 10 章 简易 LCD 示波器的设计

```c
//
    _write(0x90,15,disdata1);                    //初始开机画面,写液晶第 2 行
    _write(0x88,15,disdata2);                    //初始开机画面,写液晶第 3 行
    delay1ms(12000);                             //约 1 s
//
    setup();                                     //液晶上电初始化
    STATE = STA_Wave;                            //显示波形功能
    write_com(0x34);                             //显示波形功能
    write_com(0x36);                             //显示波形功能
    clrgdram();                                  //清屏幕
    //light = 1;                                 //关背光灯
    P1ASF = 0x03 ;                               //P1.0、P1.1 为 ADC 输入口
    ADC_POW_ON();                                //ADC 电源开
    ADC_CONTR = 0xE0;                            //选 P1.0 口 A/D 转换
//
    TMOD = 0x11;                                 //2 个定时器设定为 16 位定时器
    TH1 = 0x3C;TL1 = 0xB0;                       //50 μs 初值
    TH0 = 0xFF;TL0 = T_val;                      //采样间隔初值(T_val 初值约为 500 μs)
    ET1 = 1;TR1 = 1;
    ET0 = 1;TR0 = 1;EA = 1;
//
// ***************以下为主程序 *************************
    while(1)
    {
        readkey();
        switch(STATE){
            case STA_Info: Show_INFO();break;    //显示信息功能
            case STA_Wave: draw_bo();  break;    //显示波形功能
            case STA_Logo:             break;    //显示画面功能
            default:                   break;
        }
    }
}
//
// ******************波形定时采集中断程序 ********************
timer0() interrupt 1
{
    uchar tmp_y;
    TH0 = 0xFF;
    TL0 = T_val;                                 //采样周期可控
    draw_rigol ++ ;                              //画时间
    if(draw_rigol > = 0 && STATE == STA_Wave && draw_bo_is_ok == 0){
```

```c
                draw_rigol = 0;
                //采集波形,共 128 个点
                if(X_TIMES < 128){
                    ADC_CHANG();                    //A/D 转换
                    tmp_y = ADC_RES_DATA / Margin;
                    if(tmp_y > 63) tmp_y = 63;
                    if(tmp_y < 0) tmp_y = 0;
                    LAST_Y[X_TIMES] = NOW_Y[X_TIMES];
                    NOW_Y[X_TIMES] = tmp_y;
                    if(X_TIMES == 127) draw_bo_is_ok = 1;
                    X_TIMES ++ ;
                }
            else{
                    bo_delay ++ ;
                    if(bo_delay > 9000){
                        bo_delay = 0;
                        X_TIMES = 0;
                        //clrgdram();
                    }
                }
        }
}
//
// *************** 按键中断扫描 **********************
void timer1(void) interrupt 3
{
    TH1 = 0x3C; TL1 = 0xB0;                //50 ms 初值
    con05s ++ ;
    if(con05s >= 10)                       //0.5 s 读按键一次
    {   con05s = 0;
        if(key_done == 1)
        {
            if((P3&0xF0) != 0xF0) {        //若有键按下
                MSG_KEY = P3&0xF0;         //记录按键值
                key_done = 0;              //按键需处理标志
            }
        }
    }
}
//
/ *************** 结束 *******************/
```

第 11 章 远程电话控制器的设计

11.1 系统功能

远程电话控制器接入普通的电话线后,可以在外地任何地方用手机或座机通过拨号的方式,对连接到控制器的电器进行电源开关等操作。要求拨号接通后先进行密码核对,若3次密码错误或一定时间内没有操作则自动挂机。

11.2 设计方案

远程电话控制器利用现有公用电话网实现电器设备的远程控制,无须改造线路,安装方便,成本较低,而且由于控制器并联于普通电话机的接线上,不会影响正常电话机的使用,具有较好的应用价值。远程电话控制系统主要包括电话振铃检测电路、电话自动模拟摘挂机电路、双音多频 DTMF(Dual Tone Multi-Frequency)信号解码电路、语音提示电路、驱动接口电路和单片机控制电路等。

图 11.1 所示为远程电话控制器系统结构框架图,当通过异地的电话机拨通控制器所连接外线的电话号码时,通过市局电话交换机向用户电话机发出振铃信号,控

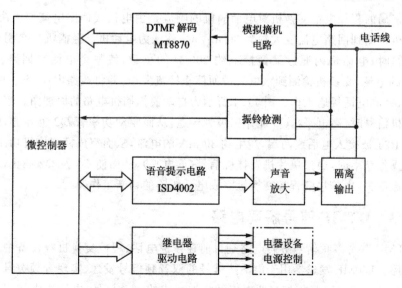

图 11.1 远程电话控制器系统结构框图

器中的振铃检测电路将检测到的振铃信号送到单片机控制器进行计数,如果控制器检测到振铃5次(可设定)后无人接听电话,则控制器自动模拟摘机,先对电话拨号者进行语音提示,若电话拨号者要求遥控操作时,则需要输入密码,接着控制器等待操作者输入密码并进行密码的校对。DTMF解码芯片将用户输入的按键信号转换成相应的数字信号,由单片机进行密码的比对,密码输入正确后再通过语音提示,使用户通过操作数字键选择控制电器及对应的电源开关等,完成后语音提示挂机结束。

11.3 系统硬件电路的设计

远程电话控制电路主要由电话振铃检测电路、自动模拟摘挂机电路、DTMF信号解码电路、语音提示电路、单片机控制电路、接口驱动电路和电源电路等组成。远程电话控制器电路原理图如11.2所示。

11.3.1 电话振铃检测电路

当用户电话被呼叫时,电信局的电话交换机发出铃流信号,振铃信号为(25 ± 3)Hz的正弦波,电压有效值约90 V,振铃重复周期为5 s,其中,1 s振铃,4 s间隙。由于振铃信号电压比较高,输入时经电容隔离降压并经全桥整流后,输入至光耦4N35的输入端,从而使光耦输出端得到低电平信号,用于单片机对铃声信号的检测和响铃计数。电话振铃检测电路原理图如图11.3所示。

11.3.2 自动模拟摘机电路

电信局的程控电话交换机对电话摘机的响应是由电话线回路电流变大决定的,当交换机检测到回路电流变大约为30 mA时,就认为电话机已经摘机。在用户摘机(接电话)时,电话机内通过叉簧接上约200 Ω的负载,使整个电话线回路流过约30 mA的电流,交换机检测到该电流后便停止铃流发送,并将线路电压变为10 V的直流电,完成电话接通工作。图11.4所示为自动模拟摘机电路的原理图。当单片机发出摘机信号时(高电平),Q3晶体三极管导通,从而驱动功率管Q2也导通,模拟负载电阻R24被接入电话线两端,产生约30 mA的电流,从而完成模拟摘机功能;当远程设定操作结束时,通过单片机发挂机信号(低电平),从而使Q3及Q2截止,电话线回路电流消失,电信局的程控交换机完成电话线路的切断工作。

11.3.3 DTMF信号解码电路

DTMF双音多频信号解码电路是目前在按键电话、程控交换机等设备中广泛应用的电路。DTMF发送器用于电话按键号的双音频信号发送,实现音频拨号。双音多频信号是一组由高频信号与低频信号叠加而成的组合信号,电话机中使用的按键与双音多频信号的频率对应关系如表11.1所列。

第 11 章 远程电话控制器的设计

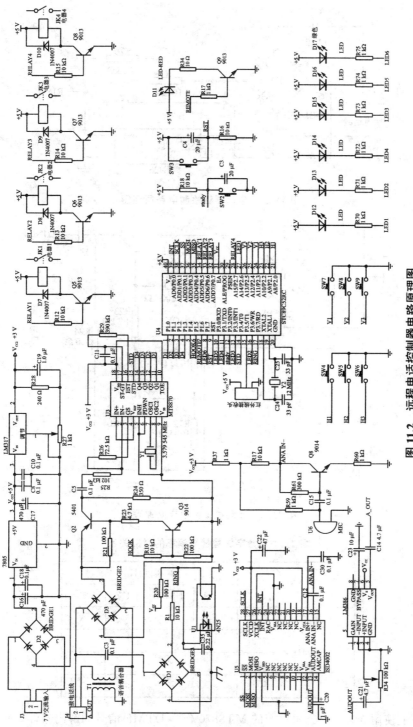

图 11.2 远程电话控制器电路原理图

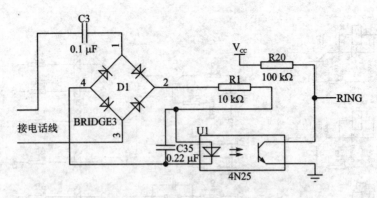

图11.3　电话振铃检测电路原理图

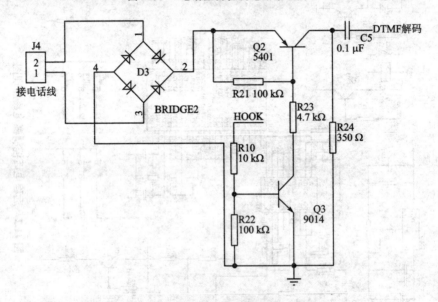

图11.4　自动模拟摘机电路原理图

表11.1　数字拨号键与频率的对应关系

数字键盘		高频组/Hz			
		1 209	1 336	1 477	1 633
低频组/Hz	697	1	2	3	A
	770	4	5	6	B
	852	7	8	9	C
	941	*	0	#	D

一般使用 MITEL 公司生产的 MT8870 作为 DTMF 信号的解码器件。MT8870 具有电路简单、功能强、功耗低、工作稳定可靠等优点。图 11.5 所示为 MT8870 解

码电路原理图。

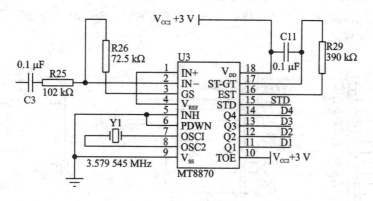

图 11.5　MT8870 解码电路原理图

当电话线上送来的双音多频信号经 C3 和 R25 送到 MT8870 的第 2 引脚（IN—）时，双音多频脉冲信号先经 MT8870 内部的拨号音滤波器，滤除拨号音信号，然后经前置放大后送入双音频滤波器，将双音频信号按高、低音频信号分开，再经高、低通滤波器以及幅度检测器后送入输出译码电路，经过数字运算后，在其数据输出端口（第 11～14 引脚）输出相对应的二进制 BCD 码。MT8870 的数据输出端 D1～D4 连到单片机 P1 口的 P1.0～P1.3 引脚，单片机经 P1 口的低 4 位识别对应的按键号。表 11.1 中的 A、B、C、D 键常被用作重拨、录放、免提等功能使用。使用解码数据（D1～D4）时要注意，对于 0 号键，MT8870 输出的二进制 BCD 码并非是 0000，而是 1010。另外，*与#键的输出分别为 1011 和 1100。当 MT8870 新获取有效双音多频信号并解码成功后，其第 15 引脚的电平由低变高，可以通知单片机取数据了，第 15 引脚的高电平在延时一段时间后会自动恢复为低电平；而无效的双音频信号或电话线路杂音、人的语音信号等是不会引起 MT8870 的数据解码的。

11.3.4　语音提示电路

远程电话控制器系统利用语音提示电路实现用户和系统的交流。语音提示电路预先存储若干段系统提示音，单片机程序根据用户发送的按键指令（DTMF 信号），进行相应的控制操作并播放语音提示下一步的操作，而操作者则根据反馈的语音信息按需要进行按键操作。

ISD4002-120P 是美国 ISD 公司系列单片语音录放集成电路的一种，采用直接模拟量存储技术，将每个采样值直接存储在片内的 Flash RAM 中，能较好地保留模拟量中的有效成分。语音芯片设计成与微处理器配合使用的寻址和控制方式，使器件引出端数减到最少，且音质较好，目前在语音合成设计中应用很广泛。ISD4002-120P 可录放 120 s 语音信号，最多可分 600 段录音，每一段均可寻址，其地址范围为 000H～258H。单片机通过模拟 SPI 接口向 ISD4002 发送上电、指定地址、开始录音、结束录音、按地址放音等操作命令,实现录音与放音操作。远程电话控制器在使用前需要提前将要提示的语音信息按段录入到芯片上。ISD4002 语音提示电路原理图如图 11.6 所示。

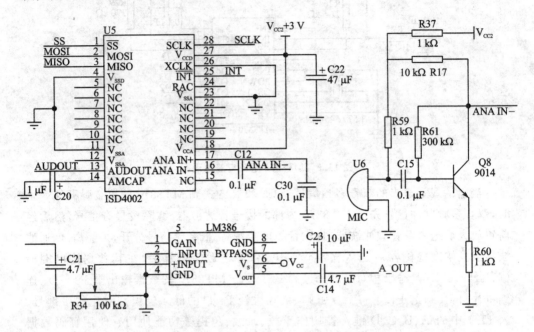

图 11.6 ISD4002 语音提示电路原理图

在录音操作时,通过麦克风将声音信号转化为电信号,并通过三极管 Q8 放大,耦合到语音信号的输入端(第 16 引脚)。单端输入时,一般信号幅度不要超过 32 mV。

在放音操作时,从 ISD4002 的第 13 引脚输出的声音信号经运放 LM386 放大后,再经过隔直耦合器送到电话线上。提示音的大小可以通过 R34 进行调整。

11.3.5 单片机控制电路

单片机采用宏晶公司的 STC89C52 系列,其具有 6 种时钟运行模式,比普通单片机的速度提高 1 倍,内带的 EEPROM 可用于密码数据的存储与修改。其主要引脚的使用如图 11.7 所示,其中 P1.0～P1.3 引脚用于双音频解码器输出数据的读入;P1.5 引脚用于模拟摘机或挂机的控制操作;P1.6 引脚输出遥控码(本实例中没有设计程序);P1.7、P3.0、P3.1、P3.3、P3.6、P2.6 引脚接 LED 指示发光二极管;第 9 引脚接上电复位电路;P3.2 引脚接遥控器学习操作键(本实例中没有设计程序);P3.4 引脚接双音频解码器的第 15 引脚,用于测试是否有双音频解码数据完成输出;P3.5 引脚为遥控码学习状态下的红外接收头信号输入口;P2.0～P2.5 引脚用于控制按键;P2.7、P0.5～P0.7 引脚用于电器的电源开关控制;P0.0～P0.4 引脚用于语音芯片 ISD4002 的录放音控制。

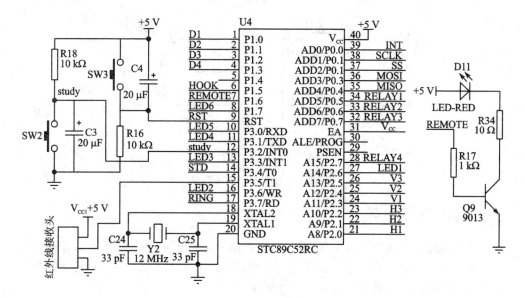

图 11.7　单片机控制电路

11.3.6　接口驱动电路

1. 电器控制接口驱动电路

电器的控制采用继电器进行交流电源的开关控制,驱动电路用小功率三极管 9013,继电器用 5 V 的直流小封装继电器,控制线圈上并联二极管用于消除反峰电压。对于不能用简单电源进行控制的电器,要用红外遥控码进行操作。本设计实例电路图中设计了红外遥控电路,但在程序中无红外线控制的内容。图 11.8 所示为继电器驱动电路图。

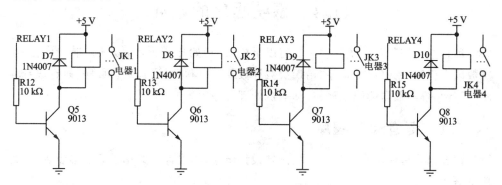

图 11.8　继电器驱动电路图

2. 按键接口驱动电路

按键设计采用最简单的顺序查键方法,6 个按键使用 P2 口的 6 根引脚线。程序中除使用录音及放音测试键外,另 4 个备用。按键接口驱动电路如图 11.9 所示。

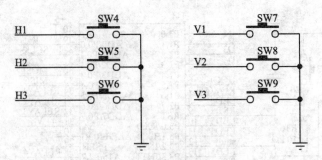

图 11.9　按键接口驱动电路

11.3.7　电源电路

电源电路需要提供两组电压,单片机系统使用 5 V 电源,语音芯片及双音频解码芯片使用 3 V 电源。设计中使用三端稳压集成电路 LM7805 及 LM317,其外围电路简单,性能稳定,电路原理图如图 11.10 所示。

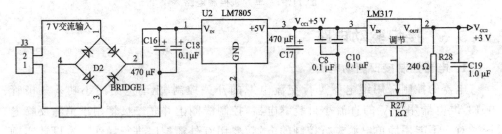

图 11.10　电源电路

11.4　系统程序的设计

远程电话控制器系统软件主要有以下模块:初始化子程序、键盘扫描程序、振铃次数读入程序、密码比较程序、错误次数计数及计时程序、语音录音程序、语音放音程序、双音频解码数据读码程序、自动摘机处理程序等。

11.4.1　语音录音程序

语音录音程序用在远程电话控制器设计阶段,本设计中共用了 5 段语音提示,每段语音时间为 10~20 s。

第 1 段内容为"欢迎来到远程家电控制系统,请输入你的密码,以 ♯ 号键结束",用于响铃 5 次后自动摘机时对操作者的提示。

第 2 段内容为"控制空调请按 1,控制电扇请按 2,控制电饭煲请按 3,控制热水器请按 4,退出请按 ✱ 号键",用于实时对远程电器的操作。

第 3 段内容为"你好,你的输入控制已完成,请挂机,谢谢",用于正常结束或 1 min 无操作动作时的提醒。

第 4 段内容为"对不起,你的密码输入有误,请重新输入密码,你有 3 次密码输入机会",用于密码输入错误的时候。

第 5 段内容为"对不起,你的输入次数已到,系统将在 10 s 后自动挂机,再见",用于密码输入 3 次错误时的挂机提醒。

设计中可根据需要录 12 段时长约为 10 s 的语音提醒语句。

ISD4002 芯片录音操作程序流程图如图 11.11 所示。

11.4.2 语音放音程序

语音放音程序用于远程自动电话操作过程中的语音提醒,不同阶段的情况要求播放对应内容的提示语音,在程序设计时要根据录音时的地址进行相应的控制并按录音的延时时间进行放音延时。ISD4002 语音放音程序流程图如图 11.12 所示。

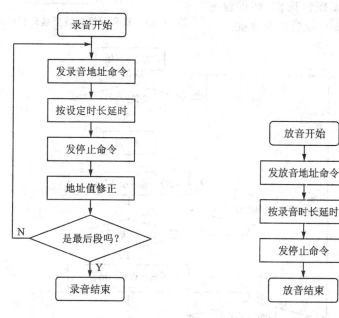

图 11.11 ISD4002 芯片录音操作程序流程图　　图 11.12 ISD4002 语音放音程序流程图

11.4.3 双音频解码数据读码程序

远程电话控制器在拨号响铃 5 次后进入自动摘机状态,先要求操作者进行密码校对,操作者使用电话机上的按键(0~9、*、♯)输入 5 位密码,并以♯号键作为结束标志;这些按键以双音频的模式传送后,被远程电话控制器中的解码芯片 MT8870

解码,当一个新的代表键号的 4 位数据解码成功后,其第 15 引脚会出现一个高电平信号,表示收到新的数据,单片机通过测试该端口判断是否要读入新的数据。读码程序的流程图如图 11.13 所示。

11.4.4 自动摘机处理程序

图 11.14 所示为远程电话控制器自动摘机处理程序流程图。在主程序循环中通过检测响铃的次数来自动摘机,摘机后提示先输入密码;在这里有 3 次输入密码的机会,然后提示控制操作的键号数字,当想结束操作时可输入 * 号键,如果在 1 min 内无按键操作,则远程电话控制器会在语音提示后自动挂机。

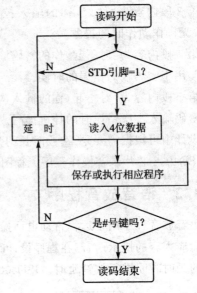

图 11.13 双音频解码数据读码程序流程图

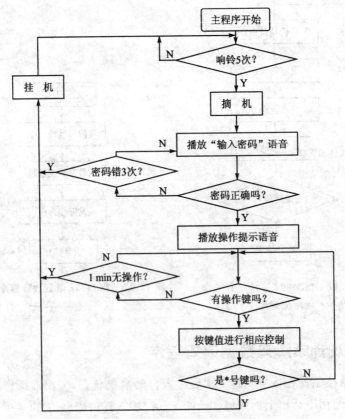

图 11.14 远程电话控制器自动摘机处理程序流程图

11.5 调试及性能分析

系统调试可分为以下 4 步:

① 调试 ISD4002 芯片的录放功能。在确保硬件电路连接正确后,先测试底座电压是否正常,单片机为 5 V 供电,语音芯片为 3 V 供电。电压正常后,插入芯片,然后调试录放程序。在录音过程中,可以通过示波器测试语音芯片的 RAC 端,正常状态下,会有高电平的时间为 175 ms、低电平的时间为 25 ms 的周期信号。若能检测到此信号,则说明芯片已经开始工作。放音时可先接扬声器进行试听,待接入电话线一切正常后再去掉扬声器。

② 语音耦合到电话线上的调试。通过电话间的相互通话,看 ISD4002 的语音是否能在另一电话机上听到。

③ 铃声计数及双音频解码电路的调试。可用示波器测试相应的脉冲波以确认电路是否正常工作。

④ 摘挂机程序的调试。由于电话遥控距离的远近与传送时间有关,所以在远距离(如人在省外或国外)操作时,按键操作的时间延时会长一些,密码输入时每次输入的时间间隔在 0.5 s 以上,控制操作的时间间隔在 1 s 以上。

利用电话远程控制系统可以实现对远距离的电话终端进行一些操作控制,不受地域的限制,操作简单且成本低。该系统也可应用于工农业生产中,实现对无人值守设备的远程控制。

11.6 源程序清单

```
/*--------------------------------------
              phone_remote control  program
              MCU STC89C52RC  XTAL 12 MHz
------------------------------------*/
// 远程电话遥控程序
//#pragma   src
#include "reg52.h"
//
#define uchar unsigned char
#define uint unsigned int
#define ulong unsigned long
#define DTMF_data P1;         //P1 口输入双音频数据(低 4 位)
#define NUM_RING 5;           //定义 5 次响铃后自动摘机
//
/*--------------------------------------
```

端口定义
-- */
//
```c
sbit LED1 = P2^6;              //摘机指示灯,红色:亮表示处于接听状态,不亮表示处于
                               //挂机状态
sbit LED2 = P3^6;              //指示灯,红色
sbit LED3 = P3^3;              //指示灯,红色
sbit LED4 = P3^1;              //指示灯,红色
sbit LED5 = P3^0;              //指示灯,红色为录音指示
sbit LED6 = P1^7;              //指示灯,绿色为放音指示
//
sbit RELAY1 = P0^5;            //继电器1
sbit RELAY2 = P0^6;            //继电器2
sbit RELAY3 = P0^7;            //继电器3
sbit RELAY4 = P2^7;            //继电器4
//
sbit BUTTON1 = P2^0;           //录音开关
sbit BUTTON2 = P2^1;           //放音测试
sbit BUTTON3 = P2^2;           //备用
sbit BUTTON4 = P2^3;           //备用
sbit BUTTON5 = P2^4;           //备用
sbit BUTTON6 = P2^5;           //备用
//
sbit HOOK_phone = P1^5;        //摘机/挂机控制口:1表示摘机,0表示挂机
sbit RING = P3^7;              //铃声输入:0表示有铃声,1表示正常
sbit STD = P3^4;               //双音频解码完成测试口(从0变成1表示完成)
sbit REMOTE = P1^6;            //遥控码输出
//ISD4002控制口
sbit SCK = P0^1;               //ISD4002串行时钟,由单片机提供
sbit ISD4002_SS = P0^2;        //ISD4002器件选择
sbit MOSI = P0^3;              //ISD4002串行输入端
sbit MISO = P0^4;              //ISD4002串行输出端
/*
sbit ISD4002_INT = P0^0;       //ISD4002中断输出
sbit ISD4002_RAC = P1^4;       //ISD4002行地址时钟输出
*/
//数据表及内存定义
uchar  data   pass_con = 0;     //密码输入个数计数
uint   data   con_1min = 0;     //1 min计数器
uchar  data   con_pulse = 0;    //响铃声次数计数器
uchar  data   con_P_W_in = 0;   //密码输入次数计数
uchar  xdata  P_W_TAB[11];      //密码数据存放
```

```c
uchar code      P_W[10] = {1,2,3,4,5,6,7,8,9,0};     //预设密码值,可修改
uchar code      phone_key_data[12] = {1,2,3,4,5,6,7,8,9,10,11,12};  //双音频解码值
//                              1 2 3 4 5 6 7 8 9 0 * #    //代表的键号
//
uchar bdata     FLASH_FLAG = 0x00;         //位标志用
sbit    out_time_FLG = FLASH_FLAG^0;       //操作超时标志(1 min)
sbit    P_W_OK = FLASH_FLAG^1;             //密码比较成功标志
/* ---------------------------------------------
                    功能函数
--------------------------------------------*/
//
/************1 ms 延时函数 *************/
delay1ms(uint t)
{
    int i,j;
    for(i = 0;i<t;i++)
        for(j = 0;j<123;j++)        //12 MHz 晶振时
            ;
}
//
//
/* ---------------------------------------------
                    SPI 读/写函数
--------------------------------------------*/
uchar SPI_rw(uchar dat)
{
    uchar i;
    for (i = 0;i<8;i++)
        {
        MOSI = dat & 0x01;
        SCK = 1;
        dat>> = 1;
        dat | = (uchar)MISO<<7;
        SCK = 0;
        }
    return dat;
}

/* ---------------------------------------------
                    ISD4002 电源唤醒
--------------------------------------------*/
void ISD4002_powerup(void)
```

```c
{
    ISD4002_SS = 0;
    SPI_rw(0x20);
    ISD4002_SS = 1;
}
```

/* --
                    ISD4002 停止命令
-- */

```c
void ISD4002_stop(void)
{
    ISD4002_SS = 0;
    SPI_rw(0x30);
    ISD4002_SS = 1;
}
```

/* --
                    ISD4002 按地址录音
-- */

```c
void ISD4002_record(uint addr)
{
    ISD4002_SS = 0;
    SPI_rw((uchar)(addr & 0x00FF));
    SPI_rw(((uchar)(addr>>8) & 0x03) | 0xA0);
    ISD4002_SS = 1;
}
```

/* --
                    ISD4002 按地址放音
-- */

```c
void ISD4002_play(uint addr)
{
    ISD4002_SS = 0;
    SPI_rw((uchar)(addr & 0x00FF));
    SPI_rw(((uchar)(addr>>8) & 0x03) | 0xe0);
    ISD4002_SS = 1;
}

#define SPEECH_NUM 12        //分12段录音,ISD4002-120 为 120 s,每秒5段行地址
                             //共600段行地址
#define SPEECH_LEN 10        //每段10 s
```
/* --

```
                       ISD4002 录音程序
--------------------------------------------*/
void speech_record(void)
{
    uchar i;
    ISD4002_powerup();
    delay1ms(25);
    ISD4002_powerup();
    delay1ms(25);
    for (i = 0;i<SPEECH_NUM;i++)
      {
          while (BUTTON1);
          ISD4002_record(i*(5*SPEECH_LEN));      //发录音命令
          LED5 = 0;                               //录音指示灯开
          delay1ms(SPEECH_LEN*1000);              //录音等待
          ISD4002_stop();                         //发录音停止命令
          LED5 = 1;                               //录音指示灯关
      }
}
/*--------------------------------------------
                       ISD4002 放音程序
--------------------------------------------*/
void speech_play(void)
{
      uchar i;
      ISD4002_powerup();
      delay1ms(25);
      ISD4002_powerup();
      delay1ms(25);
      for (i = 0;i<SPEECH_NUM;i++)
        {
            while (BUTTON2);
            ISD4002_play(i*(5*SPEECH_LEN));       //发放音命令
            LED6 = 0;                              //放音指示灯开
            delay1ms(SPEECH_LEN*1000);             //放音等待
            ISD4002_stop();                        //发停止命令
            LED6 = 1;                              //放音指示灯关
        }
}
//以下为自动语音放音用程序
//第1段语音 10 s(0～50):欢迎来到远程家电控制系统,请输入你的密码,以#号键结束
//第2段语音 20 s(50～150):控制空调请按1,控制电扇请按2,控制电饭煲请按3,控制热
```

//水器请按4,退出请按*号键
//第3段语音 10 s(150~200):你好,你的输入控制已完成,请挂机,谢谢
//第4段语音 10 s(200~250):对不起,你的密码输入有误,请重新输入密码,你有3次密码
//输入机会
//第5段语音 10 s(250~300):对不起,你的输入次数已到,系统将在10 s后自动挂机,再见
//
void play_1(void)   //欢迎来到远程家电控制系统,请输入你的密码,以#号键结束
{
        ISD4002_powerup();
        delay1ms(25);
            ISD4002_play(0);   //发放音命令
            LED6 = 0;   //放音指示灯开
            delay1ms(10000);   //放音等待
            ISD4002_stop();   //发停止命令
            LED6 = 1;   //放音指示灯关
}
//
void play_2(void)
//控制空调请按1,控制电扇请按2,控制电饭煲请按3,控制热水器请按4,退出请按*号键
{
        ISD4002_powerup();
        delay1ms(25);
            ISD4002_play(50);   //发放音命令
            LED6 = 0;   //放音指示灯开
            delay1ms(10000);   //放音等待
            ISD4002_stop();   //发停止命令
            LED6 = 1;   //放音指示灯关
}
//
void play_3(void)   //你好,你的输入控制已完成,请挂机,谢谢
{
        ISD4002_powerup();
        delay1ms(25);
            ISD4002_play(150);   //发放音命令
            LED6 = 0;   //放音指示灯开
            delay1ms(10000);   //放音等待
            ISD4002_stop();   //发停止命令
            LED6 = 1;   //放音指示灯关
}
//
void play_4(void) //对不起,你的密码输入有误,请重新输入密码,你有3次密码输入机会
{

```c
        ISD4002_powerup();
        delay1ms(25);
            ISD4002_play(200);      //发放音命令
            LED6 = 0;               //放音指示灯开
            delay1ms(10000);        //放音等待
            ISD4002_stop();         //发停止命令
            LED6 = 1;               //放音指示灯关
}
//
void play_5(void)          //对不起,你的输入次数已到,系统将在10 s后自动挂机,再见
{
        ISD4002_powerup();
        delay1ms(25);
            ISD4002_play(250);      //发放音命令
            LED6 = 0;               //放音指示灯开
            delay1ms(11000);        //放音等待
            ISD4002_stop();         //发停止命令
            LED6 = 1;               //放音指示灯关
}
//以下为摘机程序
void HOOK_phone_on(void)            //电话接听/摘机状态
{
    HOOK_phone = 1;LED1 = 0;
}
//挂机程序
void HOOK_phone_off(void)           //电话挂机状态
{
    HOOK_phone = 0;LED1 = 1;
}
//
// ****************以下为摘机处理程序*************
//
void work(void)
{
out_time_FLG = 0;                   //1 min超时标志清0
HOOK_phone_on();                    //摘机
delay1ms(500);
play_1();        //语音提示:欢迎来到远程家电控制系统,请输入你的密码,以#号键结束
//
//以下为密码输入阶段程序
con_1min = 0;                       //1 min计时清0
out_time_FLG = 0;                   //1 min超时标志清0
```

```c
    P_W_OK = 0;                        //密码比较正确标志清 0
    con_P_W_in = 0;                    //密码输入次数计数
    pass_con = 0;                      //密码输入存储位置
    //
    while(out_time_FLG == 0)
    {if(STD == 1){
        P_W_TAB[pass_con] = DTMF_data;P_W_TAB[pass_con]& = 0x0F;
        if(P_W_TAB[pass_con] == 12)
            { pass_con = 0;

        if((P_W_TAB[0] == P_W[0])&&(P_W_TAB[1] == P_W[1])&&(P_W_TAB[2] == P_W[2])&&
        (P_W_TAB[3] == P_W[3])&&(P_W_TAB[4] == P_W[4]))//5 位密码比较
                {P_W_OK = 1;goto workcon;}
            else {con_P_W_in ++ ;
                if(con_P_W_in > = 3){play_5();goto endout;} //错误密码 3 次,结束
                play_4();                       //提示密码输入有误,可输入 3 次密码
                }
            }
        else {pass_con ++ ;LED5 = 0; delay1ms(500);LED5 = 1;}   //LED5:按键接收指示
            }
    }
    //以下为操作控制程序
    workcon:
    P_W_OK = 0;                        //密码比较正确标志清 0
    con_1min = 0;                      //1 min 计时清 0
    out_time_FLG = 0;                  //1 min 超时标志清 0
    play_2();                          //控制空调请按 1,控制电扇请按 2,控制电饭煲请
                                       //按 3,控制热水器请按 4,退出请按 * 号键
    while(out_time_FLG == 0)
        {if(STD == 1){
                con_1min = 0;       //有操作,每次 1 min 计时清 0
                P_W_TAB[0] = DTMF_data;P_W_TAB[0]& = 0x0f;
                if(P_W_TAB[0] == 1){RELAY1 = ~RELAY1;LED1 = ~LED1; }
                            //按键 1——开关电器 1
                if(P_W_TAB[0] == 2){RELAY2 = ~RELAY2;LED2 = ~LED2; }
                            //按键 2——开关电器 2
                if(P_W_TAB[0] == 3){RELAY3 = ~RELAY3;LED3 = ~LED3; }
                            //按键 3——开关电器 3
                if(P_W_TAB[0] == 4){RELAY4 = ~RELAY4;LED4 = ~LED4; }
                            //按键 4——开关电器 4
                if(P_W_TAB[0] == 5){LED5 = ~LED5; }
                            //按键 5——开关小灯 5(模拟控制测试)
```

```c
            if(P_W_TAB[0] == 6){LED5 = ~LED5;}
                            //按键 6——开关小灯 5(模拟控制测试)
            if(P_W_TAB[0] == 7){LED5 = ~LED5;}
                            //按键 7——开关小灯 5(模拟控制测试)
            if(P_W_TAB[0] == 8){LED5 = ~LED5;}
                            //按键 8——开关小灯 5(模拟控制测试)
            if(P_W_TAB[0] == 9){LED5 = ~LED5;}
                            //按键 9——开关小灯 5(模拟控制测试)
            if(P_W_TAB[0] == 10){LED5 = ~LED5;}
                            //按键 0——开关小灯 5(模拟控制测试)
            if(P_W_TAB[0] == 11){goto endout;}      //遇 * 键结束
            delay1ms(1000);  //延时 1 s
            }
    }
//
//以下为挂机退出程序
endout:
play_3();                          //你好,你的输入控制已完成,请挂机,
                                   //谢谢
HOOK_phone_off();                  //挂机状态
con_1min = 0;                      //1 min 计时清 0
out_time_FLG = 0;                  //1 min 超时标志清 0
ET0 = 0;TR0 = 0;                   //关定时器
delay1ms(500);
}
//
// ****************远程电话遥控操作结束 ************
//
/* -------------------------------------------
                    void main(void)
   ------------------------------------------ */
void main(void)
{
//以下为初始化程序
REMOTE = 0;                        //关遥控码输出
HOOK_phone_off();                  //挂机状态
out_time_FLG = 0;                  //超时标志
con_1min = 0;                      // 1 min 计数器
con_pulse = 0;                     //响铃声次数计数器
con_P_W_in = 0;                    //密码输入次数计数
//
TMOD = 0x11;TH0 = 0x3C;TH1 = 0x3C;TL0 = 0xB0;TL1 = 0xB0;   //50 ms 初值
```

```c
ET0 = 0;ET1 = 0;TR0 = 0;TR1 = 0;EA = 1;    //定时器中断关闭
//
//以下为测试继电器及指示灯
RELAY1 = 1;RELAY2 = 1;RELAY3 = 1;RELAY4 = 1;
LED1 = 0;LED2 = 0;LED3 = 0;LED4 = 0;LED5 = 0;LED6 = 0;
delay1ms(1000);
RELAY1 = 0;RELAY2 = 0;RELAY3 = 0;RELAY4 = 0;
LED1 = 1;LED2 = 1;LED3 = 1;LED4 = 1;LED5 = 1;LED6 = 1;
//以下为测试语音系统/摘挂机系统
/ * HOOK_phone_on();              //摘机状态
play_1();                         //欢迎来到远程家电控制系统,请输入你的密码,
                                  //以#号键结束
LED1 = 0;delay1ms(1000);LED1 = 1; //间隔
play_2();                         //控制空调请按1,控制电扇请按2,控制电饭煲请
                                  //按3,控制热水器请按4,退出请按*号键
LED1 = 0;delay1ms(1000);LED1 = 1; //间隔
play_3();                         //你好,你的输入控制已完成,请挂机,谢谢
LED1 = 0;delay1ms(1000);LED1 = 1; //间隔
play_4();                         //对不起,你的密码输入有误,请重新输入密码,你有
                                  //3次密码输入机会
LED1 = 0;delay1ms(1000);LED1 = 1; //间隔
play_5();                         //对不起,你的输入次数已到,系统将在10 s后自动
                                  //挂机,再见
LED1 = 0;delay1ms(1000);LED1 = 1; //间隔
HOOK_phone_off();                 //挂机状态 */
//
//以下为主程序/录音/放音/远程电话遥控程序
// ******************************************
while (1)
    {
    if (!BUTTON1) {speech_record();}//录音程序
    if (!BUTTON2) {speech_play();}  //放音程序
    if (!RING) {
        delay1ms(20);if(!RING)
        { if(con_pulse == 0){out_time_FLG = 0;ET0 = 1;TR0 = 1;}//1 min 计时开始
        con_pulse ++ ;delay1ms(1500);
         if(con_pulse > = 5){con_pulse = 0;con_1min = 0;work();}
        }
            }
        }
    }
// ******************************************
```

```c
//以下为定时器中断程序
//T0 中断程序
void timer0(void) interrupt 1
{
TH0 = 0x3c;TL0 = 0xB0;
con_1min ++ ;
if(con_1min > = 1200){con_1min = 0;out_time_FLG = 1;TR0 = 0;ET0 = 0;}  //1 min 到,关计
                                                                      //时器
}
//
//T1 中断程序
void timer1(void) interrupt 3
{
TH1 = 0x3C;TL1 = 0xB0;
}
//
```

/* --
                        end
-- */

第 12 章 液晶 GPS 定位信息显示器的设计

12.1 系统功能

本设计要求利用单片机、液晶显示器和 GPS 的 OEM 板设计开发一种 GPS 定位信息显示器,要求能显示经纬度、时间和水平面高度等实时信息。

12.2 设计方案

本设计主要从 GPS 处理模块的选择、显示器的选择和 CPU 的选择 3 个方面来分析论证。

12.2.1 GPS 模块的选择

GPS 模块主要有以下几个性能指标。

1. 卫星轨迹

GPS 卫星有 24 颗,沿 6 条轨道绕地球运行(每 4 个一组),GPS 接收模块就是靠接收这些卫星来进行定位的。但是,一般在地球的同一边不会超过 12 颗卫星,所以一般选择可以跟踪 12 颗卫星以下的器件。当然,能跟踪的卫星数越多,其性能越好。大多数 GPS 接收器可以追踪 8~12 颗卫星。计算 LAT/LONG(二维)坐标至少需要 3 颗卫星,4 颗卫星可以计算三维坐标。

2. 并行通道

一般消费类 GPS 设备有 2~5 条并行通道接收卫星信号。因为最多可能有 12 颗卫星是可见的(平均值是 8 颗),GPS 接收器必须按顺序访问每一颗卫星来获取每颗卫星的信息,所以市面上的 GPS 接收器大多数是 12 并行通道型的,这允许它们可以连续追踪每一颗卫星的信息。12 通道接收器的优点包括快速冷启动和初始化卫星的信息,而且在森林地区可以有更好的接收效果。一般 12 通道接收器不需要外置天线,除非是在封闭的空间中,如船舱、车厢中。

3. 定位时间

定位时间是指重启 GPS 接收器时确定现在位置所需的时间。对于 12 通道接收器,如果在最后一次定位位置的附近,则冷启动时的定位时间一般为 3~5 min,热启动时为 15~30 s;而对于 2 通道接收器,冷启动时大多超过 15 min,热启动时为 2~5 min。

4. 定位精度

在 SA 没有开启的情况下,普通 GPS 接收器的水平位置定位精度为 5～10 m。

5. DGPS 功能

为了将 SA 和大气层折射带来的影响降为最低,有一种叫作 DGPS 发送机的设备。它是一种固定的 GPS 接收器(在 GPS 模块使用现场 100～200 km 的半径内设置),用于接收卫星的信号。DGPS 可以确切地知道理论上卫星信号传送到的精确时间是多少,然后将它与实际传送时间相比较,并计算出差值。这十分接近于 SA 和大气层折射的影响,DGPS 将这个差值发送出去,其他 GPS 接收器就可以利用这个差值得到一个更精确的位置读数(5～10 m 或者更小的误差)。许多 GPS 设备提供商在一些地区设置了 DGPS 发送机,供客户免费使用,只要客户所购买的 GPS 接收器有 DGPS 功能即可。

6. 信号干扰

要想获得一个很好的定位信号,GPS 接收器必须至少能接收 3～5 颗卫星信号。如果在峡谷中或两边高楼林立的街道上,或者在茂密的丛林里,有可能不能接收到足够的卫星信号,则无法定位或者只能得到二维坐标。同样,如果在一个建筑物里面,那么也可能无法更新位置。一些 GPS 接收器有单独的天线可以贴在挡风玻璃上,或者一个外置天线可以放在车顶上,这有助于接收器收到更多的卫星信号。

7. 其他物理指标(如大小、质量)

本设计采用的是 GARMIN 公司生产的 GPS25 - LP 型 GPS-OEM 接收板。GARMIN 公司作为全球最大的 GPS-OEM 板供应商,其生产的 GPS25 - LP 型 GPS-OEM 板具有以下主要性能指标:

- 并行 12 通道接收;
- 重捕时间小于 2 s,热启动时间为 15 s,冷启动时间为 45 s,自动搜索时间为 90 s;
- 差分(DGPS)情况下定位精度小于 5 m,非差分情况下为 15 m;
- 提供外接天线,以帮助接收;
- 体积小,功耗低,采用 5 V 供电。

12.2.2 显示器的选择

一般嵌入式系统可供选择的显示器有以下 3 种,下面将分别进行介绍。

1. VFD 显示器

VFD 显示器是由电子管发展过来的一种显示器件,它是真空二极管或三极管的一种改型。二极管的改型称为静态 VFD,三极管的改型称为动态 VFD。静态 VFD 含有两个基本电极——阴极(灯丝)和阳极,动态比静态多一极——栅极。所有极在高真空条件下封装于玻璃壳内。由阴极发射的电子在正向电位的作用下加速到达栅极和阳极(静态 VFD 直接到达阳极),并碰撞激活在阳极上的荧光粉图案使其发光。所需的亮度图形显示可以由控制栅极和阳极(静态 VFD 仅控阳极)正电位或负电位来实现。

VFD 显示器具有清晰度高、亮度高、视角宽、反应速度快及从红色到蓝色多种色彩等特点,显示效果好。当使用 CIG(集成芯片玻璃)技术时,可集成 VFD 的驱动电路,具有可靠且使用寿命长等特点,但它需要 5.5~6.3 V 的灯丝电压、150~450 mA 的灯丝电流、12~36 V 的阳极加速电压和 15~36 V 的栅极电压。不考虑阳极和栅极电流,单灯丝功耗就达 825 mW,功耗相对来说较大,不适合在移动设备上使用。另外,它需要多组电压不同的供电电压,使用不方便。

2. LED 显示器

LED 显示器是由 LED 发光二极管发展过来的一种显示器件,是 LED 发光二极管的改型,一般分为 LED 数码管显示器和 LED 点阵显示器。它具有亮度高、视角宽、反应速度快、可靠性高、使用寿命长等特点,但 LED 数码管只能显示数字和极少数的英文字符,显示单调。而 LED 点阵显示器虽然能显示各种信息,但它的体积较大,在市场上能买到的最小的 8×8 LED 点阵显示器的尺寸都有 3 cm×3 cm,适合于广告牌等需要大面积显示的地方,不适合移动设备。况且动态扫描有可能同时被点亮,此时按每段 10 mA 电流来算也有 80 mA,如果同时点亮段数更多,则电流更大。

3. LCD 液晶显示器

LCD 液晶显示器是利用光的偏振现象来显示的。一般也分为数字型 LCD(同 LED 数码管显示器,只能显示数字和极少数的英文字符)和点阵型 LCD 两种:前者用于只须显示简单字符的地方,如时钟等;后者能显示各种复杂的图形和自定义的字符,因此应用比较广泛。LCD 液晶器具有本身不发光,靠反射或者透射其他光源的优点,同时具有功耗小、可靠性高、寿命长(工业级>100 000 h,民用级>50 000 h)、体积小、电源简单等特点,非常适合于嵌入式系统、移动设备和掌上设备的使用。

本设计采用点阵型 LCD 液晶显示器 CGM-12232。该显示器具有 122×32 点阵,不仅可以显示数字,还可以显示中文、英文甚至图片等,体积只有 61 mm×19 mm×5.7 mm,功耗仅为 5 V×2.5 mA=12.5 mW(不开背光)。

12.2.3 CPU 的选择

一般 GPS 导航器都是 GPS 配合矢量电子地图来进行导航和航线记录。这些设备 CPU 的运算量和需要储存的数据量都很大,一般都使用 X86、ARM 等 32 位 CPU。考虑到本设计只须显示经纬度和时间等简单信息,决定选择 ATMEL 公司的 AT89C52 单片机作为主控制器。

12.3 系统硬件电路的设计

图 12.1 所示为 GPS 定位信息显示器系统设计原理框图。

第 12 章 液晶 GPS 定位信息显示器的设计

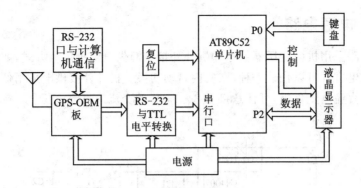

图 12.1 GPS 定位信息显示器系统设计原理框图

系统硬件电路主要由 GPS-OEM 接收板、液晶显示器、AT89C52 单片机、键盘、RS-232 电平转换、单片机上电复位和电源等部分组成。GPS-OEM 板发送的串行数据经 RS-232（CMOS/TTL 电平转换）电路送至单片机串行口，经处理后通过键盘选择要显示的信息，送至 LCD 液晶显示器。LCD 液晶显示器为定时更新，更新周期约为 1 s。上电复位电路为单片机上电提供上电复位。电源电路为各个电路提供稳定的 +5 V 电源。GPS-OEM 板的设置用预留的 RS-232 口，在计算机上用 GARMIN 公司提供的软件（GARMIN Sensor/Smart Antenna Software）来进行设置。

图 12.2 所示为 GPS 定位信息显示器电路原理图。

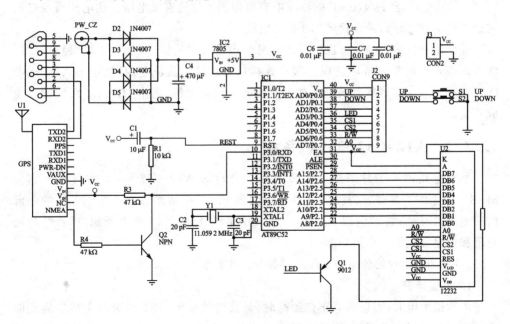

图 12.2 GPS 定位信息显示器电路原理图

12.3.1 电源电路

电源电路采用机内变压器供电和机外外接电源供电两种供电方式,如图12.3所示。两种供电方式可以任选一种,在机内自动切换。机外外接供电采用傻瓜式接口,不需要辨认直流电正负极,可任意接入8~15 V的交流或直流电压。

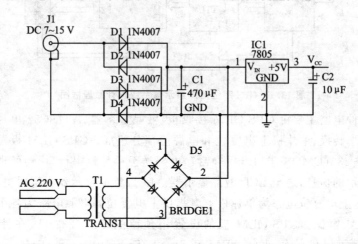

图12.3 电源电路原理图

机内变压器输入220 V交流电压,输出7 V交流电压。经过桥式整流输出大约9 V的脉动电压,经过470 μF的滤波电容可得到平稳的直流电压。此电压再经过三端稳压器7805稳压,输出稳定的+5 V电压。

外接供电口输入的电源也经过机内另一组桥式整流桥,再经过滤波、稳压、然后输出。输入口的整流桥堆实现了傻瓜式接口。当输入直流电源时,由D1、D4或者D2、D3中的一组完成极性转换。如果输入的是交流电源,则由D1~D4组成的桥完成整流。

12.3.2 AT89C52单片机系统

系统电路的主芯片采用美国ATMEL公司的AT89C52 Flash单片机。它与MCS-51系统产品兼容,具有4 KB可重编程Flash存储器,5(1±20%)V的电源使用电压,128×8位的内部RAM,2个16位定时/计数器,6个中断源,以及低功耗空闲和掉电方式等一系列功能。

AT89C52单片机的电源、复位、晶振振荡电路如图12.4所示。

1. 复位电路

单片机上电时,当振荡器正在运行时,只要持续给出RST引脚两个机器周期的高电平,便可完成系统复位。外部复位电路是为内部复位电路提供两个机器周期以上的高电平而设计的。系统采用上电自动复位,上电瞬间电容器上的电压不能突变,

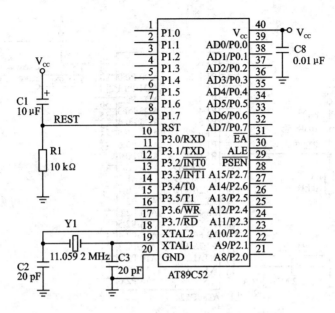

图 12.4　AT89C52 单片机的电源、复位、晶振振荡电路图

RST 上的电压是 V_{CC} 上的电压与电容器上的电压之差,因而 RST 上的电压与 V_{CC} 上的电压相同。随着充电的进行,电容器上的电压不断上升,RST 上的电压就随着下降,RST 上只要保持 10 ms 以上的高电平,系统就会有效复位。电容 C1 可取 10～33 μF,电阻 R1 可取 1.2～10 kΩ。在系统设计中,C1 取 10 μF,R1 取 10 kΩ,充电时间常数为 $10\times10^{-6}\times10\times10^{3}=100$ ms。

2. 晶振振荡电路

XTAL1 引脚和 XTAL2 引脚分别构成片内振荡器的反相放大器的输入和输出端,外接石英晶体或陶瓷振荡器以及补偿电容 C1、C2 构成并联谐振电路。当外接石英晶体时,电容 C2、C3 选 30 pF±10 pF;当外接陶瓷振荡器时,电容 C2、C3 选 47 pF±10 pF。AT89C52 系统中晶振可在 0～24 MHz 选择。外接电容 C2、C3 的大小会影响振荡器频率的高低、振荡频率的稳定度、起振时间及温度稳定性。

在设计电路板时,晶振和电容应靠近单片机芯片,以减少寄生电容,保证振荡器稳定可靠工作。

在系统设计中,为保证串行通信波特率的误差,选择了 11.059 2 MHz 的标准石英晶振,电容 C2、C3 为 20 pF。

12.3.3　键盘电路

键盘电路原理图如图 12.5 所示。

本键盘为最简单的点式键盘,由单片机 I/O 口进行扫描。一般来说,键盘按键多数采用行列式,如图 12.6 所示。这是因为在按键数量多时行列式键盘在占用相同

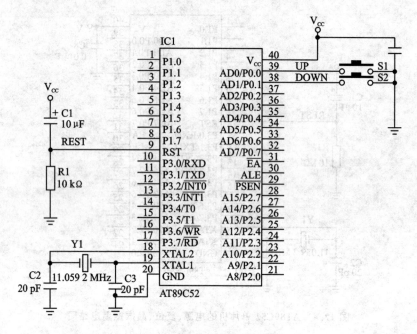

图 12.5　键盘电路原理图

数量 I/O 口时,能设置的按键较点式键盘多。但在按键少时行列式键盘还不如点式键盘来得简单、方便。本设计只设置两个按键,用来进行显示信息的翻页。

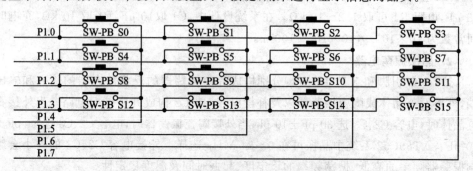

图 12.6　典型的 4×4 行列式键盘图

12.3.4　单片机与 GPS-OEM 板接口电路

GARMIN GPS25-LP 型 GPS-OEM 板输出引脚功能如图 12.7 所示。

GPS-OEM 板各引脚功能介绍如下:

引脚 1:串行口 2 的数据输出端。

引脚 2:串行口 2 的数据输入端。

引脚 3:秒脉冲输出端,精度为 ±10 μs。

引脚 4:串行口 1 的数据输出端。

第 12 章 液晶 GPS 定位信息显示器的设计

引脚 5：串行口 1 的数据输入端。
引脚 6：掉电模式控制端。
引脚 7：外部备用电源输入端。
引脚 8：GND 接地端。
引脚 9：V_{IN} 电源输入端。
引脚 10：同引脚 9，电源输入端，内部与引脚 9 相连。
引脚 11：空脚 NC。
引脚 12：NMEA（美国海洋电子协会）0183，Ver1.5 格式语句输出端。

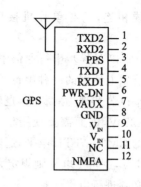

图 12.7 GPS-OEM 板输出引脚图

由于使用的是 LVS 版本 GPS25 – LP 型 GPS-OEM 板，所以串行口 1、串行口 2 和 NMEA 口使用的都是 RS – 232 电平。如果使用的是 LVC 版本的 GPS25 – LP 型 GPS-OEM 板，则端口是 CMOS/TTL 电平。在本系统中，将串行口 2 用于计算机作 GPS-OEM 板设置用，本机显示 GPS 信息从 NMEA 口送出。GPS-OEM 板接口电路如图 12.8 所示。

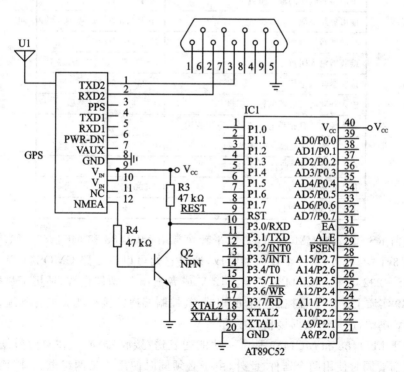

图 12.8 GPS-OEM 板接口电路

由于 GPS-OEM 板送出的是 RS – 232 电平，计算机串行通信用的也是 RS – 232 电平，单片机使用的是 CMOS/TTL 电平，因此 GPS-OEM 板与计算机通信可以直接用串行线相连,而与单片机接口必须进行 RS-232 电平和 CMOS/TTL 电平的转换。

RS-232 是异步串行通信中应用最早的,也是最广泛的标准串行总线之一;它原是基于公用电话网的一种串行通信标准,推荐电缆的长度最长为 15 m(50 ft);它的逻辑电平以公共地为对称,其逻辑 0 电平规定为 +3~+25 V,逻辑 1 电平则规定为 -3~-25 V,因而需要使用正负极性的双电源。其主要电气参数如表 12.1 所列。而传统的 CMOS/TTL 电平,逻辑电平是以地为标准不对称设置,其逻辑 0 电平规定小于 0.7 V,逻辑 1 电平则规定大于 3.2 V。因此,两者之间的逻辑电平不兼容,两者通信时必须进行电平转换。

表 12.1 RS-232 标准的主要电气参数表

项 目	参数指标
带 3~7 kΩ 负载时的驱动器输出电平	逻辑 0 为 +3~+25 V, 逻辑 1 为 -3~-25 V
不带负载时的驱动器输出电平	-25~+25 V
驱动器通电时的输出阻抗	大于 300 Ω
输出短路电流	小于 0.5 A
驱动器转换速率	小于 30 V/μs
接收器输入阻抗	3~7 kΩ
接收器输入电压	-25~+25 V
输入开路时接收器的输出逻辑	1
输入经 300 Ω 接地时接收器的输出逻辑	1
+3 V 输入时接收器的输出逻辑电平	0
-3 V 输入时接收器的输出逻辑电平	1
最大负载电容	2 500 pF
不能识别的过渡区	-3~+3 V

以前 RS-232 与 CMOS/TTL 电平转换常用 MC1488 和 MC1489。MC1488 实质上由 3 个"与非"门和 1 个反向器组成,通过它们可以将 4 路 CMOS/TTL 电平转换为 RS-232 电平。MC1488 需要 ±12 V 或者 ±15 V 双路电源,适用于数据发送。MC1489 实质上是 4 个带控制门的反相器,其控制端通常接一滤波电容到地,使用单一 +5 V 电源,适用于数据接收。

由于 MC1488 和 MC1489 是单一功能的发送/接收器,所以在双向数据传输中各端都需要同时使用两个器件;此外,由于必须同时使用正负两组电源,因而在很多场合下使用显得不方便,所以被淘汰。为此推出了只用单一电压且具有发送/接收双重功能的电路。这种器件内部集成了一个电荷泵和一个电压变换器,它能将 +5 V 或者更低的电压变换成 RS-232 所需的 ±10 V 以上的电压。这类芯片最典型的就是 MAXIM 的 MAX232 芯片,如图 12.9 所示。其内部电荷泵电路先将 +5 V 提升到 +10 V,然后再用电压反转电路将 +10 V 变成 −10 V,这样就得到了 RS-232 所需的 ±10 V 的电压了。

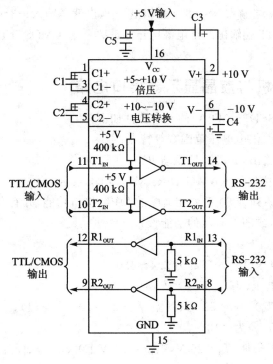

图 12.9 MAX232 接线图

本设计单片机只须接收从 GPS-OEM 板发送过来的数据,而无须向 GPS-OEM 板发送数据。也就是说,只须将 RS-232 电平转换为 CMOS/TTL 电平,而无须将 CMOS/TTL 电平转换为 RS-232 电平。通过对 RS-232 电平和 CMOS/TTL 电平(见表 12.2)分析,决定采用单个三极管来进行转换,而不用 MAX232 等专用 RS-232 与 CMOS/TTL 电平转换器,具体接线如图 12.10 所示。

表 12.2 RS-232 电平和 CMOS/TTL 电平逻辑电平对比表

逻辑值	RS-232 电平/V	CMOS/TTL 电平/V
0	+3~+25	小于 0.7
1	−3~−25	大于 3.2

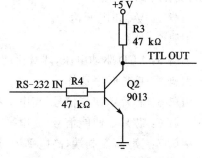

图 12.10 RS-232 电平 CMOS/TTL 电平转换图

当 RS-232 IN 端输入 RS-232 逻辑电平 0,也就是输入+3~+25 V 时,三极管正向导通,此时 TTL OUT 端输出的是三极管的饱和压降。此电压为 0.1~0.2 V,符合 CMOS/TTL 电平小于 0.7 V 的要求。

当 RS-232 IN 端输入 RS-232 逻辑电平 1,也就是输入-3~-25 V 时,三极管截止,此时 TTL OUT 端输出的是电源电压+5 V,符合 CMOS/TTL 电平大于 3.2 V 的要求。

12.3.5 单片机与液晶显示器接口电路

液晶驱动 IC(SED1520F0A)的基本特性如下:
- 具有功耗低、供电电压范围宽等特点;
- 具有 61 段输出,并可外接驱动 IC 扩展驱动;
- 具有 2560 位显示 RAM(DD RAM),即 80×8×4 位;
- 具有与 68 系列或 80 系列相适配的 MPU 接口功能,并有专用的指令集,可完成文本显示或图形显示的功能设置。

工作参数介绍如下:
- 逻辑工作电压($V_{DD}-V_{SS}$)为 2.4~6.0 V;
- LCD 驱动电压($V_{DD}-V_{LCD}$)为 3.0~13.5 V;
- 工作温度(T_a)为 0~55 ℃(常温)/ -20~70 ℃(宽温);
- 保存温度(T_{stg})为 -10~70 ℃。

电气特性(测试条件: $T_a=25$ V,$V_{DD}=5.0$ V±0.25 V)如下:
- 输入高电平(V_{IH})为 3.5 V(min);
- 输入低电平(V_{IL})为 0.55 V(max);
- 输出高电平(V_{OH})为 3.75 V(min);
- 输出低电平(V_{OL})为 1.0 V(max);
- 工作电流为 2.0 mA(max)。

液晶显示器 CGM-12232 的引脚功能如图 12.11 所示,具体介绍如下:

引脚 1:V_{DD} 正电源输入。
引脚 2:地。
引脚 3:LCD 驱动电压,调对比度。
引脚 4:接口时序选择。
引脚 5:芯片 1 使能信号,高电平有效。
引脚 6:芯片 2 使能信号,高电平有效。
引脚 7:读/写控制端,高电平读,低电平写。
引脚 8:数据/指令选择端,高电平为数据,低电平为指令。

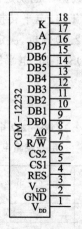

图 12.11 液晶显示器 CGM-12232 的引脚功能图

第12章 液晶GPS定位信息显示器的设计

引脚9～16：液晶并行数据。

引脚17：背光LED阳极。

引脚18：背光LED阴极。

单片机与液晶显示器接口电路图如图12.12所示。CGM-12232的引脚9～16接单片机P2口进行数据传输，引脚5～8的3根控制线接P0口。由于P0口内部没有上拉电阻不能输出高电平，因此，在P0口上接了一个10 kΩ排阻RP9作为P0口的上拉电阻。LCD液晶显示器的背光LED灯采用三极管驱动控制。

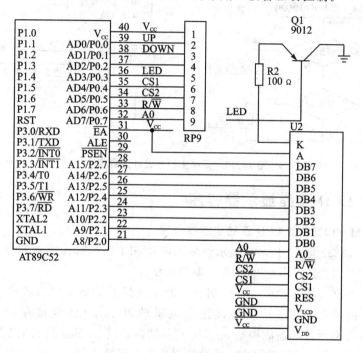

图12.12　单片机与液晶显示器接口电路图

在CGM-12232的说明文档中，引脚V_{LCD}须通过电阻在V_{DD}与GND之间分压得到，但实验中发现，通过分压后液晶显示很暗，而直接将其接地会使液晶显示明显好转，因此这里将其直接接地。

12.4　系统程序的设计

系统软件采用C语言编写，C编译器为Keil C51 7.10版本，文本编辑环境和编译环境为MedWin。

12.4.1　系统软件设计原理

系统软件运行总体设计流程如下：系统初始化，显示开机画面，串行中断接收

GPS-OEM板的"＄GPGGA"语句,每当正确收到"＄GPGGA"语句一次,就更新一次显示,键盘可以选择显示的GPS信息。

系统主程序设计流程图如图12.13所示。

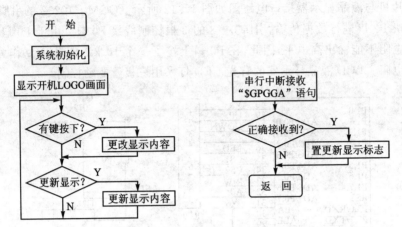

图 12.13　系统主程序设计流程图

12.4.2　LCD 液晶显示器程序

1. CGM－12232 型 LCD 液晶显示器原理

CGM－12232型LCD液晶显示器采用两片SED1520F0A驱动芯片,LCD液晶显示程序即是对两片SED1520F0A的驱动程序。

SED1520F0A属行列驱动及控制合一的小规模液晶显示驱动芯片,电路简单,经济实用,内含振荡器,只须外接振荡电阻即可工作(已经接在板上)。一个SED1520F0A显示控制器能控制 $80×16$ 点阵液晶的显示,其显示RAM共16行,分2页,每页8行,每一页的数据寄存器分别对应液晶屏幕上的8行点。当设置了页地址和列地址后,就确定了显示RAM中的唯一单元。屏幕上的每一列对应一个显示RAM的字节内容,且每列最下面一位为MSB,最上面一位为LSB,即该RAM单元字节数据由低位到高位的各个数据位对应于显示屏上某一列的由高到低的8个数据位。对显示RAM的一个字节单元赋值就是对当前列的8行(一页)像素点是否显示进行控制。

CGM－12232液晶显示器的引脚定义如表12.3所列。

各符号含义如下:

● DB0～DB7:数据总线。

● A0:数据/指令选择信号。A0＝1,表示出现在数据总线上的是数据;A0＝0,表示出现在数据总线上的是指令或读出的状态。

● RES:接口时序类型选择。RES＝0为Intel 8080时序(见图12.14和表12.4),操作信号是$\overline{RD}$和$\overline{WR}$;RES＝1为M6800时序(见图12.15和表12.5),其操

第12章 液晶GPS定位信息显示器的设计

作信号是 $\overline{CS}$ 和 $R/\overline{W}$。

- $\overline{RD}(\overline{CS})$：在 Intel 8080 时序时为读,低电平有效;在 M6800 时序时为使能信号,是个正脉冲,在低电平时为写操作,在高电平时为读操作。
- $\overline{WR}(R/\overline{W})$：在 Intel 8080 时序时为写,低电平有效;在 M6800 时序时为读、写选择信号,$R/\overline{W}=1$ 为读,$R/\overline{W}=0$ 为写。

表 12.3 CGM-12232 液晶显示器的引脚定义表

序号	符号	状态	功能说明	序号	符号	状态	功能说明
1	V_{DD}	—	逻辑电源正	7	$R/\overline{W}$	输入	读/写选择信号
2	GND	—	逻辑电源地	8	A0	输入	寄存器选择信号
3	V_{LCD}	—	液晶显示驱动电源	9~16	DB0~DB7	三态	数据总线
4	RES	—	接口时序类型选择	17	A	—	背光灯正电源
5	CS1	输入	主工作方式IC使能信号	18	K	—	背光灯负电源
6	CS2	输入	从工作方式IC使能信号				

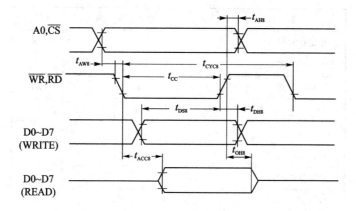

图 12.14 与 Intel 8080 系列单片机接口读/写操作时序图

表 12.4 与 Intel 8080 系列单片机接口时序参数表($V_{DD}=5.0(1\pm10\%)$ V,$T_a=-20\sim+75$ ℃)

名称	符号	最小值	最大值	单位	名称	符号	最小值	最大值	单位
地址建立时间	t_{AW8}	20		ns	数据建立时间	t_{DS8}	80		ns
地址保持时间	t_{AH8}	10		ns	写数据保持时间	t_{DH8}	10		ns
$R/\overline{W},\overline{WR}$周期	t_{CYC8}	1000		ns	读存取时间	t_{ACC8}		90	ns
$R/\overline{W},\overline{WR}$脉冲宽度	t_{CC}	200		ns	读数据保持时间	t_{CH8}	10	60	ns

SED1520F0A 液晶显示驱动器有13条指令。表 12.6 所列为以 M6800 系列单片机接口为例(RES=1)总结出的指令。

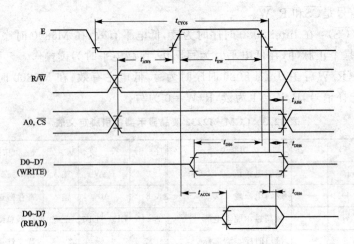

图 12.15 与 M6800 系列单片机接口读/写操作时序图

表 12.5 与 M6800 系列单片机接口时序参数表
($V_{DD}=5.0(1\pm10\%)$ V, $T_a=-20\sim+75$ ℃)

名 称		符 号	最小值	最大值	单 位
地址建立时间		t_{AW6}	20		ns
地址保持时间		t_{AH6}	10		ns
系统时钟周期		t_{CYC6}	1000		ns
脉冲宽度	读	t_{EW}	100		ns
	写		80		ns
数据建立时间		t_{DS6}	80		ns
写数据保持时间		t_{DH6}	10		ns
存取时间		t_{ACC6}		90	ns
读数据保持时间		t_{OH6}	10	60	ns

表 12.6 SED1520F0A 指令集

指令名称	控制信号		控制代码							
	$R/\overline{W}$	A0	D7	D6	D5	D4	D3	D2	D1	D0
显示开/关指令	0	0	1	0	1	0	1	1	1	I/O
显示起始行设置	0	0	1	1	0	显示起行(0~31)				
设置页地址	0	0	1	0	1	1	1	0	页地址(0~3)	
设置列地址	0	0	0	列地址(0~79)						
读状态指令	1	0	BUSY	ADC	OFF/ON	RESET	0	0	0	0

续表 12.6

指令名称	控制信号		控制代码							
	R/$\overline{W}$	A0	D7	D6	D5	D4	D3	D2	D1	D0
写数据	0	1	显示的数据							
读数据	1	1	显示的数据							
ADC 选择指令	0	0	1	0	1	0	0	0	0	0/1
静态驱动开/关	0	0	1	0	1	0	0	1	0	0/1
占空比选择	0	0	1	0	1	0	1	0	0	0/1
改写开始指令	0	0	1	1	1	0	0	0	0	0
改写结束指令	0	0	1	1	1	0	1	1	1	0
复位	0	0	1	1	1	0	0	0	1	0

SED1520F0A 指令集功能介绍如下:

- 显示开/关指令功能:开/关屏幕显示,不改变显示 RAM(DD RAM)中的内容,也不影响内部状态。D=0,开显示;D=1,关显示。如果在显示关闭的状态下选择静态驱动模式,那么内部电路将处于安全模式,以减小功耗。安全模式下的内部状态:停止 LCD 驱动 Segment 和 Common 输出 V_{DD} 电平;停止晶体振荡并禁止外部时钟输入,晶振输入 OSC2 引脚处于不确定状态;显示数据和内部模式不变。
- 显示起始行设置指令功能:执行该命令后,所设置的行将显示在屏幕的第一行。起始地址可以是 0~31 范围内任意一行。行地址计数器具有循环计数功能,用于显示行扫描同步,当扫描完一行后自动加 1,直到 31。
- 设置页地址指令功能:设置页地址。当单片机要对 DD RAM 进行读/写操作时,首先要设置页地址(见表 12.7)和列地址。本指令不影响显示 RAM(DD RAM)中的内容。
- 设置列地址指令功能:设置 DD RAM 中的列地址。当单片机要对 DD RAM 进行读/写操作前,首先要设置页地址和列地址(见表 12.8)。执行读写命令后,列地址会自动加 1,直到达到 50H 才会停止,但页地址不变。

表 12.7 页地址对应表

A1	A0	页地址
0	0	0
0	1	1
1	0	2
1	1	3

表 12.8 列地址对应表

A6	A5	A4	A3	A2	A1	A0	列地址
0	0	0	0	0	0	0	0
0	0	0	0	0	0	1	1
			⋮				
1	0	0	1	1	1	0	4E
1	0	0	1	1	1	1	4F

- 读状态指令功能：检测内部状态。
  - BUSY 为忙信号位，BUSY=1，内部正在执行操作；BUSY=0，空闲状态。
  - ADC 为显示方向位，ADC=0，反向显示；ADC=1，正向显示。
  - ON/OFF 为显示开关状态，ON/OFF=0，显示打开；ON/OFF=1，显示关闭。
  - RESET 为复位状态，RESET=0，正常；RESET=1，内部正处于复位初始化状态。
- 写数据指令功能：将 8 位数据写入 DD RAM。该指令执行后，列地址自动加 1，所以可以连续将数据写入 DD RAM，而不用重新设置列地址。
- 读数据指令功能：读出页地址和列地址限定的 DD RAM 地址内的数据。当"读—修改—写"模式关闭时，每执行一次读指令，列地址自动加 1，所以可以连续从 DD RAM 读出数据而不用设置列地址。注意：在设置完列地址后，首次读显示数据前必须执行一次空的"读显示数据"，这是因为设置完列地址后，第 1 次读数据时，出现在数据总线上的数据是列地址，而不是所要读出的数据。
- ADC 选择指令功能：静态驱动开/关指令功能。D=0 表示关闭静态显示，D=1 表示打开静态显示。如果在打开静态显示时，执行关闭显示指令，则内部电路将被置为安全模式。
- 占空比选择指令功能：D=0 表示 1/16 DUTY，D=1 表示 1/32 DUTY。
- 改写开始指令功能：执行该指令后，每执行一次写数据指令，列地址自动加 1；但执行读数据指令时，列地址不会改变。这个状态一直持续到执行 END 指令。注意：在"读—修改—写"模式下，除列地址设置指令之外，其他指令照常执行。
- 改写结束指令功能：关闭"读—修改—写"模式，并把列地址指针恢复到打开"读—修改—写"模式前的位置。
- 复位指令功能：使模块内部初始化。复位指令对显示 RAM 没有影响。初始化内容包括设置显示初始行为第 1 行，页地址设置为第 3 页。

2. CGM-12232 型 LCD 液晶显示器软件设计

CGM-12232 型 LCD 液晶显示器软件设计从底层写起，逐步提高，最后完成显示一个 code 区域的数据功能。即先从往 LCD 液晶显示器发送 1 字节的数据或指令写起，逐步上升，最后到画一个图指定开始列、上下层、图形的宽度、图形指针固定高度为 16 的图。汉字和数字字符都是通过"画"这种图来得到的。

(1) 往 LCD 液晶显示器发送 1 字节的数据或指令子程序

程序原型如下：

调用方式：void send_mi(uchar instuction)

函数说明：发指令 instruction 到主窗口（内函数，私有，用户不能直接调用）。

调用方式：void send_si(uchar instruction)

函数说明：发指令 instruction 到从窗口（内函数，私有，用户不能直接调用）。

第 12 章 液晶 GPS 定位信息显示器的设计

调用方式：void OutMD(uchar i)
函数说明：发数据 data 到主窗口（内函数，私有，用户不能直接调用）。
调用方式：void send_sd(uchar data)
函数说明：发数据 data 到从窗口（内函数，私有，用户不直接调用）。

首先要确定往主芯片，还是从芯片发送，然后判断芯片是否忙，忙则等待，不忙则可以发数据或者指令，最后把选中的芯片取消选中。其流程图如图 12.16 所示。

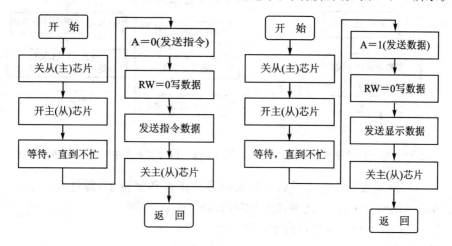

图 12.16　LCD 液晶指令（左）、数据（右）发送流程图

（2）芯片判忙子程序

程序原型如下：

调用方式：void wait_ready(void)
函数说明：等待 LCD 内部操作完成，判忙（内函数，私有，用户不能直接调用）。

芯片判忙是本系统唯一读 LCD 液晶显示器的一个子程序。读取当前 LCD 的状态，以判断 LCD 是否忙，用到的是读状态指令。芯片判忙子程序流程如图 12.17 所示。

读出数据和 0x80 相"与"，直到结果为 0 为止（不忙），也就是等待读状态指令里面读出的 BUSY 位（D7）为 0 为止，说明芯片不忙。

（3）LCD 液晶屏初始化子程序

程序原型如下：

调用方式：void lcd_init(void)
函数说明：12232 LCD 液晶屏初始化，开机后仅调用一次。

仅开机时调用一次，主要负责设置 LCD 液晶屏的一些状态，包括主芯片复位，从芯片复位；关主芯片显示，关从芯片显示；设置主芯片动态显示，设置从芯片动态显示；设置主芯片 1/32 占空比，设置从芯片 1/32 占空比；设置主芯片时钟线模式，设置从芯片时钟线模式；主芯片改写指令结束，从芯片改写指令结束；设置主芯片起始行

0行起始列0列,设置从芯片起始行0行起始列0列;液晶初始化结束返回。具体流程如图12.18所示。

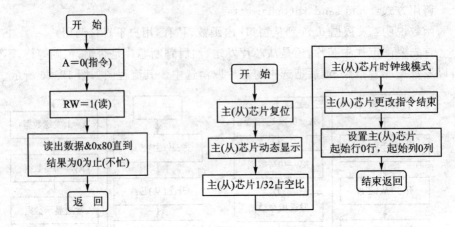

图12.17　芯片判忙子程序流程图　　图12.18　LCD液晶屏初始化子程序流程图

LCD液晶屏初始化完成后就可以显示各种图形和字符了,即进入正常工作状态。在显示图形和字符时,还要注意及时切换页(page)和列。

(4) 页切换子程序和列切换子程序

程序原型如下:

调用方式:void set_page(uchar page)

函数说明:同时设置主(右)从(左)显示页为0～3页。

调用方式:void SetAddress(uchar address)

函数说明:同时设置主(右)从(左)列地址为0～61列。

页切换子程序主要是在4个页面之间进行切换,列切换子程序主要是设置起始列。它们两个都是发送一个特定的数据指令给LCD液晶屏。页切换发送的是1011 10××,后两位××表示页地址;列切换子程序发送的是0××× ××××,后几位××× ××××表示起始列的地址。由于这两个程序只是简单地将两个特定值发送给LCD液晶屏,所以对于其行地址这里就不再详述。

(5) 清屏子程序

程序原型如下:

调用方式:void lcd_clr(void)

函数说明:清屏。

起始清屏子程序是在整个画面上画一个空白的图片。SED1520F0A有点类似于Flash ROM芯片,但刚好与Flash ROM相反。Flash ROM置0是只要把0写进去即可,置1则需要擦除。可以在原来1的基础上把1变成0,而不可以把原来的0变成1。而SED1520F0A可以把原来的0变成1,而不可以把原来的1变成0,也就是说,刚好与Flash ROM相反。因此,在每一次要重新显示画面时必须调用清屏程序。清屏子程序流程如图 12.19 所示。

(6) 画图子程序

程序原型如下：

调用方式：void draw_bmp(uchar col,uchar layer,uchar width,uchar * bmp)

函数说明：画一个图，横坐标是 col，layer 表示上下层，width 是图形的宽，高固定为 16，bmp 是图形数据指针；使用 Zimo3Pro 软件，采用纵向取模，字节倒序得到数据。

　　col：图形的起始位置 0～121。

　　layer：图形的位置(Y 坐标)。0 表示上半部分，非 0 表示下半部分。

　　width：图形宽度要求 col+width<121。

画图子程序是 LCD 初始化完成后对 LCD 液晶显示器唯一的操作函数，是操作 LCD 液晶显示器的基础。画图子程序流程如图 12.20 所示。

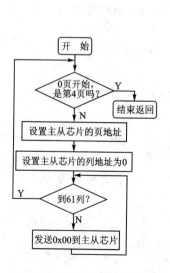

图 12.19　清屏子程序流程图

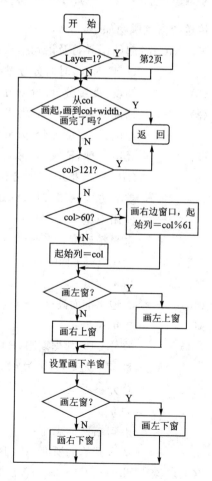

图 12.20　画图子程序流程图

12.4.3 GPS 接收子程序

　　GPS 接收子程序主要用于接收 GPS25-LVS 板发送的串行数据。这个程序在串行中断里面完成。GPS25-LVS 的通信波特率默认值为 4 800,1 个起始位,8 个数据位,1 个停止位,无奇偶校验。通常使用 NMEA-0183 格式输出,数据代码为 ASCII 码字符。NMEA-0183 是美国海洋电子协会为海用电子设备制定的标准格式,目前广泛使用 V2.0 版本。由于该格式为 ASCII 码字符串,比较直观且易于处理,所以在许多高级语言中都可以直接进行判别、分离,以提取用户所需要的数据。GPS25-LVS 系列 OEM 板可输出 12 句语句,分别是 GPGGA、GPGSA、GPGSV、GPRMC、GPVTG、LCGLL、LCVTG、PGRME、PGRMF、PGRMT、PGRMV 和 GPGLL。这里接收的是"$GPGGA"这条语句的数据。"$GPGGA"语句的格式为

$GPGGA,〈1〉,〈2〉,〈3〉,〈4〉,〈5〉,〈6〉,〈7〉,〈8〉,〈9〉,M,〈10〉,M,〈11〉,〈12〉*hh〈CR〉〈LF〉

传送的信息说明如下:
$GPGGA　　起始引导符及语句格式说明(本句为 GPS 定位数据);
〈1〉　　　　UTC 时间,时时分分秒秒格式;
〈2〉　　　　纬度,度度分分.分分分分格式(第 1 位是 0 也将传送);
〈3〉　　　　纬度半球,N 或 S(北纬或南纬);
〈4〉　　　　经度,度度度分分.分分分分格式(第 1 位是 0 也将传送);
〈5〉　　　　经度半球,E 或 W(东经或西经);
〈6〉　　　　GPS 质量指示,0 为方位无法使用,1 为非差分 GPS 获得方位,2 为差分方式获得方位(DGPS),6 为估计获得;
〈7〉　　　　使用卫星数量,为 00~12(第 1 个是 0 也将传送);
〈8〉　　　　水平精确度,0.5~99.9;
〈9〉　　　　天线离海平面的高度,-9999.9~9999.9 m;
M　　　　　指单位 m;
〈10〉　　　 大地水准面高度,-999.9~9999.9 m;
M　　　　　指单位 m;
〈11〉　　　 差分 GPS 数据期限(RTCM SC-104),最后设立 RTCM 传送的秒数量(若无 DGPS 则为 0);
〈12〉　　　 差分参考基站标号,从 0000~1023(首位 0 也将传送,若无 DGPS 则为 0);
*　　　　　语句结束标志符;
hh　　　　　从 $ 开始的所有 ASCII 码的校验和;
〈CR〉　　　 此项在 GPS25-LVS 板中不传送;
〈LF〉　　　 此项在 GPS25-LVS 板中不传送。

实时收到的一条"＄GPGGA"语句如下：

＄GPGGA,114641,3002.3232,N,12206.1157,E,1,03,12.9,53.2,M,11.6,M,,*4A

这是一条 GPS 定位数据信息语句，意思为 UTC 时间为 11 时 46 分 41 秒，位置在北纬30°2.3232′,东经 122°6.1157′,普通 GPS 定位方式，接收到 3 颗卫星，水平精度为 12.9 m，天线离海平面高度为 53.2 m，所在地离地平面高度为 11.6 m，校验和为 4AH。

由于 GPS-OEM 板发送的不止一条语句，且要完整地接收这条＄GPGGA 语句，就必须判断这条语句的头，也就是"＄GPGGA,"这 7 个字符，当完整地收到这 7 个字符后，才能保证是所需要的数据。具体流程如图 12.21 所示。

12.4.4 键盘子程序

键盘子程序的用途是根据键盘输入更改显示的 GPS 定位信息，同时亮 LED 背光灯 10 s。可以选择显示的信息包括经纬度、北京时间和大地水准面高度 3 条信息。这里安排了两个按键：up 和 down 键。但由于这里只显示 3 条信息，所以硬件连接时只接了一个，而软件仍旧按两个来写，以备以后扩展时使用。具体流程如图 12.22 所示。

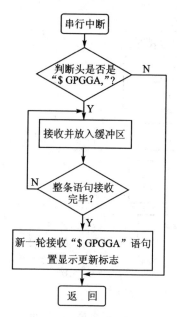

图 12.21 单片机串行中断接收子程序流程图

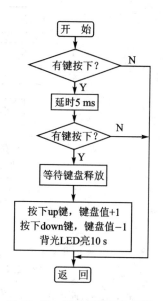

图 12.22 键盘子程序流程图

12.4.5 显示子程序

显示子程序是根据键盘值将经纬度、北京时间和大地水准面高度这 3 条信息选择一个显示到 LCD 液晶显示器。由于 GPS-OEM 板发送的是 ASCII 码的数据,所以在显示前必须将 ASCII 码转成 BCD 码;而对于 0~9 以外的字符则根据需要转到特定值,然后根据键盘值显示所要显示的内容。

1. 显示经纬度

对应经度的格式为"度度分分.分分分分,E(W)"。

对应纬度的格式为"度度分分.分分分分,N(S)"。

显示的信息为

> 东经:×××°××′××″
> 北纬:××°××′××″

其中,GPS-OEM 板发送的信息和要显示的信息有所不同。首先,必须将 GPS-OEM 板发送的"E,S,W,N"对应成"东,南,西,北"4 个字符;其次,GPS-OEM 板发送的是"度度分分.分分分分"格式,而要显示的是"度度°分分′秒秒″"的格式,因此必须进行转换。转换的方法是×100/60。另外,考虑到 GPS-OEM 板最后的两位"分分"即使在完全不动的情况下变化也会没有规律,因此把这两位舍去。具体流程如图 12.23 所示。

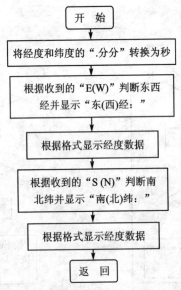

图 12.23 经纬度显示程序流程图

其中:经度先显示 3 位数字,然后显示单位"°";再显示 2 位数字,再显示"′";再显示 2 位数字,再显示"″"。纬度也是一样,只不过刚开始显示的是两位数字,并且显示在 LCD 液晶显示器的下半部。

2. 显示北京时间

对应时间格式为"时时分分秒秒"。

显示的信息为

初看起来好像 GPS-OEM 板传过来的时间格式与要显示的时间格式是一致的,似乎不用转换。但实际上 GPS-OEM 板传过来的是格林尼治时间,也就是东一区的时间。北京时间与格林尼治时间相差 8 小时,即北京时间＝格林尼治时间＋8 小时。但当格林尼治时间 16 点以后,北京时间已经是第 2 天的凌晨了,也就是说,当算出来的北京时间大于 24 时,必须减去 24 才是正确的北京时间。具体流程如图 12.24 所示。

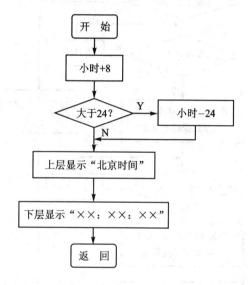

图 12.24　北京时间显示流程图

3. 显示大地水准面高度

选择大地水准面高度,而不选择天线离海平面的高度,主要是因为大地水准面高度相对稳定,而天线离海平面的高度值变化较大,在一次实验中,天线离海平面的高度值在短短半个小时内从 135.6 m 变化到 −37.4 m,而大地水准面高度基本稳定在 11.6 m。

由于大地水准面高度值必须在 GPS-OEM 接收到卫星信号,并且锁定以后才会传送,因此也只有当 GPS-OEM 接收到卫星信号,并且锁定以后才能显示。在软件中,对语句头后的第 32 位,也就是卫星接收个数进行判断,只当这个值不等于 0 时,才能显示大地水准面高度。

另外,天线离海平面的高度、大地水准面高度在串行缓冲里面的数据位置都是浮动的。这是因为这两个数据都是有几位数据就传几位数据,而到底有多少位数据要根据实际所处位置而定。因此,在显示这些信息之前必须探测这些数据的位置。大地水准面高度在天线离海平面高度的单位"M"与它自己的单位"M"之间,寻找这两个"M"的位置就可以找到大地水准面高度这个数据的位置。

大地水准面高度显示流程如图 12.25 所示。

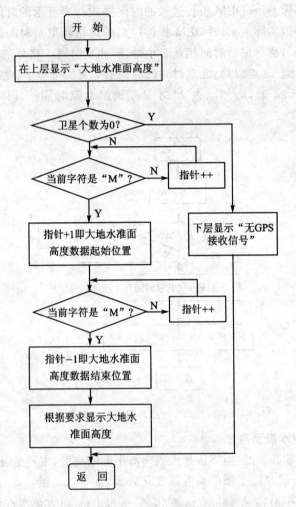

图 12.25　大地水准面高度显示流程图

12.4.6　初始化子程序

系统初始化包括 RAM 初始化、特殊功能寄存器初始化和外围设备初始化。RAM 初始化主要是将 RAM 进行清 0 处理。本系统只用到内部的 RAM,即只须对内部的 RAM 进行清 0 处理即可。特殊功能寄存器的初始化包括定时器的初始值的装入、中断的开放等。外围设备初始化主要是对外围设备的初值设定,例如本系统就在上电时必须将 LCD 液晶显示器进行初始化。

一般在 Keil C51 编译器里面 RAM 的初始化由"STARTUP.A51"文件完成,它其实是一个带有宏定义和条件编译的汇编语言文件。开机先执行这个文件产生的代码,然后再执行 main()主函数。

特殊功能寄存器初始化和外围设备初始化在 main()主函数处开始,在进入功能函数循环前完成初始化。本系统主要完成端口复位、开机延时、LCD 液晶屏初始化、LCD 液晶屏清屏、调用开机 LOGO 画面、定时器 T0 初始化、串行口波特率发生器(T1)初始化、开定时器 T0 和串行口中断。具体流程如图 12.26 所示。

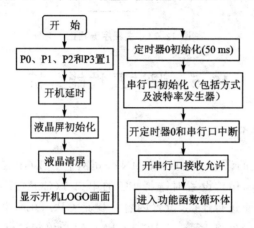

图 12.26 开机初始化流程图

开机 LOGO 画面为亮背光显示：

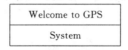

然后延时 3 s,再显示：

然后延时 1.5 s,打开定时器 T0,10 s 后熄灭背光。

定时器 T0 主要是背光显示延时 10 s 熄灭用。当按键按下时,将背光打开,然后将定时器 T0 打开,定时器 T0 定时为 50 ms,配合一个软件计数器,在 10 s 后将背光关掉,同时也将自身关闭,停止计时。

12.4.7 主程序

主程序是一个无限循环体。先扫描键盘,然后判断是否需要更新显示。若需要,则更新显示;若不需要,则返回。

具体流程如图 12.27 所示。

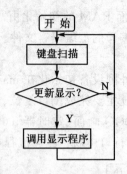

图 12.27 主程序流程图

12.5 调试及性能分析

12.5.1 调试步骤

① GPS-OEM 板的设置。从 GARMIN 公司网上下载 GPS25-LVS 系列产品应用软件(文件名为 gpscfg),设置时选择 PC 的串口 1(COM1)或串口 2(COM2)与 GPS-OEM 板的串口 1 进行连接,设置内容主要是 GPS-OEM 板的通信波特率及输出语句,本设计选择 4800 波特率,选择"$GPGGA"单语句输出。

② 硬件及软件综合调试。先进行电路板的静态测试,然后通电检测,再结合软件进行调试,最后接上 GPS 有源天线并将天线置于露天场所进行综合调试。

12.5.2 性能分析

液晶 GPS 定位信息显示系统的时间为原子钟时间,因此非常精确,而定位经纬度的误差是由 GPS-OEM 板的性能决定的,GPS25-LVS 系列方位定位精度可达 10 m 左右,能满足一般应用项目的使用。

12.6 源程序清单

以下是 GPS 信息显示器 C 程序清单:

```
/*****************************************************************/
/*                    GPS 信息显示器 C 程序                       */
/*****************************************************************/
# include <reg52.h>           //AT89C52 单片机头文件
# include <LCD_code.h>        //液晶 LCD 的字模文件
# include <intrins.h>

//功能引脚定义
```

第 12 章 液晶 GPS 定位信息显示器的设计

```
sbit A  = P0^7;              //数据1/命令0选择
sbit RW = P0^6;              //读1/写0
sbit E1 = P0^4;              //片选1(Master)
sbit E2 = P0^5;              //片选2(Slave)
sbit LED = P0^3;             //背光
sbit up = P0^0;              //向上翻页键
sbit down = P0^1;            //向下翻页键
#define data P2               //液晶并行数据
//液晶显示控制命令表
#define disp_on 0xAF          //显示关闭
#define disp_off 0xAE         //显示打开
#define disp_start_line 0xC0  //显示起始地址(后5位表示0～31行)
#define page_addr_set 0xB8    //页地址设置(0～3)
#define col_addr_set 0x00     //列地址设置(0～61)
#define status_busy 0x80      //0 = ready
#define modeRWite 0xEE        //写模式
#define dynamic_driver 0xA4   //动态驱动
#define adc_select 0xA0       //clockwise
#define clk32 0xA9            //刷新时钟设置1/32
#define clk16 0xA8            //刷新时钟设置1/16
#define reset 0xE2            //软件复位
#define uchar unsigned char
#define uint unsigned int
//全局变量,及标志位定义
uchar time_counter = 0;       //定时器的软件计数器
uchar key = 0;                //键盘值
uchar serial_counter;         //串行计数器
bit disp_flag = 0;            //显示更新标志
uchar bdata serial_byte = 0;  //串行口标志位定义字节
sbit Sflag  = serial_byte^0;  //串行接收头部标志,8个都定义在 serial_byte 内
sbit G1flag = serial_byte^1;
sbit Pflag  = serial_byte^2;
sbit G2flag = serial_byte^3;
sbit G3flag = serial_byte^4;
sbit Aflag  = serial_byte^5;
sbit DFflag = serial_byte^6;
sbit ENflag = serial_byte^7;
bit r_flag = 0;
unsigned char idata serial_buff[77];    //串行接收缓冲
void lcd_init(void);                     //LCD 初始化
```

```c
void lcd_clr(void);                //LCD 清屏
void wait_ready(void);             //等待 ready
void draw_bmp(uchar col,uchar layer,uchar width,uchar * bmp);
                                   //点阵码显示输出
void ASCII2BCD(void);              //ASCII 码转换为 BCD 码
void logo(void);                   //开机画面显示
/* - - - - - - - - - - - - - - - -
             中断程序
/* - - - - - - - - - - - - - - - */
// ***** 定时器 T0 中断函数,用于控制背光灯延时 10 s 熄灭 *****
void int_t0() interrupt 1 using 1
{                                  //定时器 T0 中断函数,用于控制背光灯延时 10 s 熄灭
  TH0   = 0x4C;
  TL0   = 0x00;                    //重装定时器 T0,定时 50 ms
  time_counter ++ ;                //软件计数器 + 1
  if (time_counter == 200)         //软件计数器定时到 10 s,关背光,定时器 T0,清软件计数器
    {
      time_counter = 0;
      LED = 1;
      TR0 = 0;
    }
}

// **** 串行口中断函数,用于语句"$ GPGGA"判断和此语句的接收 ****
void serial() interrupt 4 using 2
{
 uchar pp;
 RI = 0;
 pp = SBUF;
 if(ENflag == 1)                   //串口接收完毕,可以用来显示,清标志位重新开始
     {
         disp_flag = 1;
         serial_byte = 0;
     }
   else if(DFflag == 1)            //"$ GPGGA"头判断完毕,开始接收"$ GPGGA"语句的数据
       {
            if(pp == 42)
                ENflag = 1;  //等待收到"*"结束接收
            else
              {
                  serial_buff[serial_counter] = pp;
                                   //没收到"*",继续接收,数据放入串口缓冲
```

```c
                serial_counter ++ ;
            }
        }
    else if(Aflag == 1)            //第 6 个为"A",判断第 7 个是否为","
        {
            if(pp == 44)
                DFflag = 1;        //第 7 个是",",开始接收"$GPGGA"语句的数据
            else
                serial_byte = 0;   //不是",",清标志位
        }
    else if(G3flag == 1)           //第 5 个为"G",判断第 6 个是否为 A
        {
            if(pp == 65)
                Aflag = 1;         //第 6 个是 A,判断下一个是否为","
            else
                serial_byte = 0;   //不是 A,清标志位
        }
    else if(G2flag == 1)           //第 4 个为 G,判断第 5 个是否为 G
        {
            if(pp == 71)
                G3flag = 1;        //第 5 个是 G,判断下一个是否为 A
            else
                serial_byte = 0;   //不是 G,清标志位
        }
    else if(Pflag == 1)            //第 3 个为 P,判断第 4 个是否为 G
        {
            if(pp == 71)
                G2flag = 1;        //第 4 个是 G,判断下一个是否为 G
            else
                serial_byte = 0;   //不是 G,清标志位
        }
    else if(G1flag == 1)           //第 2 个为 G,判断第 3 个是否为 P
        {
            if(pp == 80)
                Pflag = 1;         //第 3 个是 P,判断下一个是否为 G
            else
                serial_byte = 0;   //不是 P,清标志位
        }
    else if(Sflag == 1)            //第 1 个为 $,判断第 2 个是否为 G
        {
            if(pp == 71)
                G1flag = 1;        //第 2 个是 G,判断下一个是否为 P
```

```
                    else
                        serial_byte = 0;        //不是 G,清标志位
                }
        else if(pp == 0x24)                     //判断第 1 个是否为 $
            {
                Sflag = 1;                      //第 1 个为 $,判断下一个是否为 G
                serial_counter = 0;             //串行计数器清 0
            }
}
//调用方式: void send_mi(uchar instuction)
//函数说明: 发指令 instruction 到主窗口(内函数,私有,用户不能直接调用)
void send_mi(uchar instruction)
{
E2 = 0;                                 //关 Slaver
E1 = 1;                                 //开 Master
wait_ready();                           //判断忙
A = 0;                                  //指令
RW = 0;                                 //写触发
data = instruction;                     //指令码
E1 = 0;                                 //关 Master
}

//调用方式: void OutMD(uchar i)
//函数说明: 发数据 data 到主窗口(内函数,私有,用户不能直接调用)
void send_md(uchar c)
{
E2 = 0;                                 //关 Slaver
E1 = 1;                                 //开 Master
wait_ready();                           //判断忙
A = 1;                                  //数据
RW = 0;                                 //写触发
data = c;                               //数据
E1 = 0;                                 //关 Master
}

//调用方式: void send_si(uchar instruction)
//函数说明: 发指令 instruction 到从窗口(内函数,私有,用户不能直接调用)
void send_si(uchar instruction)
{
E1 = 0;                                 //关 Master
E2 = 1;                                 //开 Slaver
wait_ready();                           //判断忙
A = 0;                                  //指令
```

第12章 液晶GPS定位信息显示器的设计

```c
    RW = 0;                         //写触发
    data = instruction;             //指令码
    E2 = 0;                         //关 Slaver
}

//调用方式: void send_sd(uchar data)
//函数说明: 发数据 data 到从窗口(内函数,私有,用户不直接调用)
void send_sd(uchar c)
{
    E1 = 0;                         //关 Master
    E2 = 1;                         //开 Slaver
    wait_ready();                   //判断忙
    A = 1;                          //数据
    RW = 0;                         //写触发
    data = c;                       //数据
    E2 = 0;                         //关 Slaver
}

//等待 ready: 等待 LCD 内部操作完成,判忙
void wait_ready(void)
{
    A = 0;                          //指令
    RW = 1;                         //读
    _nop_();                        //空操作,产生汇编里面的 NOP
    while(data & status_busy);      //读入 LCD 状态,1 = 忙,一直等待 LCD 内部操作完成
}

//调用方式: void lcd_init(void)
//函数说明: 122 × 32 LCD 初始化,开机后仅调用一次
void lcd_init(void)
{
    send_mi(reset);                 //复位 m-left,s-right
    send_si(reset);

    send_mi(disp_off);              //关闭显示
    send_si(disp_off);

    send_mi(dynamic_driver);        //动态驱动
    send_si(dynamic_driver);

    send_mi(clk32);                 //1/32 占空比
    send_si(clk32);

    send_mi(adc_select);            //顺时针
    send_si(adc_select);
```

```c
    send_mi(modeRWite);                    //写模式结束
    send_si(modeRWite);

    send_mi(col_addr_set);                 //归回零列,设定显示起始行首
    send_mi(disp_start_line);
    send_si(col_addr_set);
    send_si(disp_start_line);

    send_mi(disp_on);                      //开显示
    send_si(disp_on);
}
//调用方式: void lcd_clr(void)
//函数说明:清屏
void lcd_clr(void)
{
uchar i, page;
for (page = 0;page < 4;page ++ )
  {
     send_mi(page_addr_set|page);         //设置页从 0~3
     send_si(page_addr_set|page);
     send_mi(0);                          //主窗口设置为 0 列
     send_si(0);                          //从窗口设置为 0 列
     for (i = 0;i < 62;i ++ )             //全部写入 0x00
       {
          send_md(0x00);
          send_sd(0x00);
       }
  }
}

//调用方式: void set_page(uchar page)
//函数说明:同时设置主(右)从(左)显示页为 0~3 页
void set_page(uchar page)
{
send_mi(page_addr_set|page);
send_si(page_addr_set|page);
}

//调用方式: void SetAddress(uchar address)
//函数说明:同时设置主(右)从(左)列地址为 0~61 列
void set_address(uchar address)
{
send_mi(address&0x7F);                    //&0x7F,考虑到防止越限
send_si(address&0x7F);
```

第 12 章　液晶 GPS 定位信息显示器的设计

}

//调用方式:void putchar_l(uchar c)
//函数说明:在左页(主窗口)当前地址画 1 字节(8 点)
void putchar_l(uchar c)
{
send_md(c);
}

//调用方式:void putchar_r(uchar c)
//函数说明:在右页(从主窗口)当前地址画 1 字节(8 点)
void putchar_r(uchar c)
{
send_sd(c);
}

//调用方式: void draw_bmp(uchar col,uchar layer,uchar width,uchar * bmp)
//函数说明:画一个图,横坐标是 col,layer 表示上下层,width 是图形的宽度,高固定为 16
//bmp 为图形数据指针
//使用 Zimo3Pro 软件,采用纵向取模,字节倒序得到数据
//col 为图形的起始位置,0~121
//layer 为图形的位置(Y 坐标),0 表示上半部分,非 0 表示下半部分
//width 为图形的宽度,8、16 可选
void draw_bmp(uchar col,uchar layer,uchar width,uchar * bmp)
{
uchar x;
uchar address;   //address 表示显存的物理地址
uchar p = 0;
uchar page = 0;
uchar window = 0;   //page 表示上下两页,window 表示左右窗口(0 为左,1 为右)
if (layer) page = 2;   //左为主窗口,右为从窗口

for (x = col; x < col + width; x ++)
{
　　if (x > 121)return;   //防止显示乱码
　　if (x > 60)   //左右窗口定位
　　　{
　　　　window = 1;   //右为从窗口
　　　　address = x % 61;
　　　}
　　else
　　{
　　　address = x;
　　}   //主窗口输出

```c
    set_page(page);                    //上层数据输出
    set_address(address);
    if (window)
      {
         putchar_r(bmp[p]);
      }
    else
    {
      putchar_l(bmp[p]);
    }
    set_page(page + 1);                //下层数据输出
    set_address(address);              //列保持不变
    if (window)
       {
          putchar_r(bmp[p + width]);
       }
    else
       {
          putchar_l(bmp[p + width]);
       }
    p ++ ;
  }
}
/* ---------------------------------------
            延时程序：入口 unsinged char i ,延时 i ms
----------------------------------------*/
delay(uchar i)
{
  uchar j;
  for(i = i<<1;i>0;i--)
  for(j = 0xF3;j>0;j--)
  {}
}

// ****键盘扫描程序,有键按下,修改 key 值****
keywork(void)
{
    if(!(up&&down))                    //是否有键按下
    {
       delay(20);                      //延时 5 ms
       if(up&&down)                    //再次判断是否有键按下,没有返回
```

```
        {
          return;
         }
        if(!up)                          //查 up 键
          {
            while(!up);                  //等待 up 键释放
            key ++ ;                     //key 值加 1
            LED = 0;                     //亮背光
            TR0 = 1;                     //开定时器 T0,背光亮 10 s
            time_counter = 0;            //清定时器软件计数器,从按键以后亮背光 10 s
            if (key == 3)                //key 值超上限,置下限
              {
                key = 0;
              }
            return;
          }
        if(! down)                       //查 down 键
          {
            while(! down);               //等待 down 键释放
            key - - ;                    //key 值减 1
            LED = 0;                     //亮背光
            TR0 = 1;                     //开定时器 T0,背光亮 10 s
            time_counter = 0;            //清定时器软件计数器,从按键以后亮背光 10 s
            if (key == 255)
              {
                key = 2;                 //key 值超上限,置下限
              }
            return;
          }
     }

}
disp_transit()
{
   uchar j, row = 0 ,temp;
   temp = (serial_buff[25] * 10 + serial_buff[26]) * 3/5;
                        //经度最后".××"分转换为秒
   serial_buff[25] = temp/10;
   serial_buff[26] = temp % 10;
   temp = (serial_buff[12] * 10 + serial_buff[13]) * 3/5;  //纬度最后".××"分转换为秒
   serial_buff[12] = temp/10;
```

```c
    serial_buff[13] = temp % 10;
    lcd_clr();                                              //清屏
    draw_bmp(0,0,16,azimuth[serial_buff[30] - 20]);         //判断东、西,并显示
    draw_bmp(16,0,16,jing);                                 //显示"经"
    draw_bmp(32,0,8,num[13]);                               //显示":"
    for(j = 19;j < 28;j ++ )                                //显示经度
      {
        if (j == 22)
          {
              draw_bmp(40 + row * 8,0,8,num[11]);           //显示数字度符号
              row ++ ;
          }
        if (j == 24)
          {
              draw_bmp(40 + row * 8,0,8,num[12]);           //显示"'"(分)
              row ++ ;
              j ++ ;
          }

        if (j == 27)                                        //显示""(秒)
          {
              draw_bmp(40 + row * 8,0,8,num[14]);
              row ++ ;
              break;
          }
        draw_bmp(40 + row * 8,0,8,num[serial_buff[j]]);     //显示数字
        row ++ ;

      }

row = 0;
draw_bmp(0,1,16,azimuth[serial_buff[17] - 20]);             //判断南、北,并显示
draw_bmp(16,1,16,wei);                                      //显示纬
draw_bmp(32,1,8,num[13]);                                   //显示":"
for(j = 7;j < 15;j ++ )
  {
    if (j == 9)                                             //显示数字度符号
      {
          draw_bmp(40 + row * 8,1,8,num[11]);
          row ++ ;
      }
```

```c
        if (j == 11)
          {
             draw_bmp(40 + row * 8,1,8,num[12]);         //显示"'"(分)
             row ++ ;
             j ++ ;
        if (j == 14)                                     //显示"""(秒)
          {
             draw_bmp(40 + row * 8,1,8,num[14]);
             row ++ ;
             break;
          }
          draw_bmp(40 + row * 8,1,8,num[serial_buff[j]]);  //显示数字
          row ++ ;
        }
     }
disp_time()
{
   uchar hour, j, row = 0;
   lcd_clr();                                           //清屏
   draw_bmp(29,0,64,bjsj);                              //在液晶上部 29 列开始显示北京时间
   hour = serial_buff[0] * 10 + serial_buff[1] + 8;     //格林尼治时间转化为北京时间
   if (hour > 23)                                       //北京时间 = 格林尼治时间 + 8
   {hour = hour - 24;}                                  //当大于 24 时,减去 24
   serial_buff[0] = hour/10;                            //回存到缓冲区
   serial_buff[1] = hour % 10;
   for( j = 0;j<6;j ++ )                                //显示 6 个数字和 2 个冒号,格式为××:××:××
     {
        draw_bmp(29 + row * 8,1,8,num[serial_buff[j]]);  //显示 6 个数字
        row ++ ;
        if ((j == 1)||(j == 3))                          //第 2 个和第 4 个数字后面显示":"
          {
             draw_bmp(29 + row * 8,1,8,num[13]);         //显示冒号
             row ++ ;
          }
     }
}
disp_level()
{
 uchar i,j,row = 1;
 lcd_clr();
 draw_bmp(0,0,120,level);
```

```c
    if(serial_buff[32] == 0)                    //无 GPS 接收信号
      {
        draw_bmp(0,1,112,nosignal);
      }
    else
      {                                         //探测水平信号数据存放的位置
        for (i = 35;row;)
          {
            if (serial_buff[i] == 'M')
              {
                i = i + 2;
                row = 0;                        //跳出循环
              }
            else
              {
                i ++ ;
              }
          }
        j = i;
        row = 1;
        for (;row;)
          {
            if (serial_buff[j] == 'M')
              {
                row = 0;                        //跳出循环
              }
            else
              {
                j ++ ;
              }
          }

        for (;i<j-1;i++ )
          {
            draw_bmp(0 + row * 8,1,8,num[serial_buff[i]]);
            row ++ ;
          }
        draw_bmp(0 + row * 8,1,8,num[15]);
      }
  }

disp_speed()                                    //速度显示程序
  {
```

```
}
disp_direction()                          //方向显示程序
{

}
/* ---------------------------------------
                显示程序,根据键盘值调用不同的显示函数
----------------------------------------*/
display()
{
 ASCII2BCD();              //先调用 ASCII 码转换函数,将 ASCII 码转换为 BCD 码

 switch(key)
   {                                      //根据键盘值调用不同的显示函数
      case 0 : { disp_transit();   break;}   //键盘值 0,显示经纬度
      case 1 : { disp_time();      break;}   //键盘值 1,显示北京时间
      case 2 : { disp_level();     break;}   //键盘值 2,显示大地水准面高度
      case 3 : { disp_speed();     break;}   //键盘值 3,显示速度,未扩展
      case 4 : { disp_direction();break;}    //键盘值 4,显示方向,未扩展
      default: { disp_transit();   break;}
                                           //可扩充
   }
}


/* ---------------------------------------
                将接收到的 ASCII 码转换位 BCD 码和特定值
----------------------------------------*/
void ASCII2BCD(void)
{
  uchar i;
  for(i = 0;i<60;i++)              //只转换前 38 个接收到的数据
    {
       if((serial_buff[i] >= 48) && (serial_buff[i] <= 57))
          {                         //接收到的是数字,转换为 BCD 码。BCD 码 = ASCII 码 - 48
            serial_buff[i] = serial_buff[i] - 48;
          }
       else
          {
            switch( serial_buff[i] )
               {                    //非数字,将它们转换为特定值
                  case '.' : { serial_buff[i] = 10; break; }
```

```c
                    case '-' : { serial_buff[i] = 16; break; }

                    case 'E' : { serial_buff[i] = 20; break; }
                    case 'S' : { serial_buff[i] = 21; break; }
                    case 'W' : { serial_buff[i] = 22; break; }
                    case 'N' : { serial_buff[i] = 23; break; }
                    case ',' : { serial_buff[i] = 0xff;break;}
                    default : break;
                    }
                }
            }
        }
}
/* ---------------------------------------
                开机 LOGO 画面,开机时调用一次
----------------------------------------*/
void logo(void)
{
    uchar i, j, h;
    LED = 0;                            //点亮背光灯
    TR0 = 1;                            //开定时器 T0,背光灯亮 10 s
    lcd_clr();
    draw_bmp(0,0,112,welcome_1);        //在液晶上部第 0 列开始显示"Welcome to GPS"
    draw_bmp(32,1,48,welcome_2);        //在液晶下部第 32 列开始显示显示"System"
    for(i = 0;i<10;i++ )for(j = 1;j;j++ )for(h = 1;h;h++ );
                                        //延时 3 s
    lcd_clr();                          //清屏
    draw_bmp(0,0,64,welcome_3);         //在液晶上部第 0 列开始显示"设计者:"
    draw_bmp(64,1,48,welcome_4);        //在液晶下部第 64 列开始显示"唐召波"
    for(i = 0;i<5;i++ )for(j = 1;j;j++ )for(h = 1;h;h++ );
                                        //延时 1.5 s
}
// ***********************************************************
//                        主程序
// ***********************************************************
void main(void)
{
    P0    = 0xFF;                       //端口复位
    P1    = 0xFF;
    P2    = 0xFF;
    P3    = 0xFF;
    delay(255);
```

```
    lcd_init();                          //初始化液晶显示器
    lcd_init();
    lcd_init();
    lcd_clr();                           //清屏
    logo();                              //显示开机画面
//TMOD: GATE|C/! T|M1|M0|GATE|C/! T|M1|M0
//       0   0   1  0  0   0   0   1
    TMOD = 0x21;                         //T0 为 16 位定时器,T1 为自动重装,方式 3
    TH0  = 0x4C;
    TL0  = 0x00;                         //定时 50 ms
//SCON: SM0|SM1|SM2|REN|TB8|RB8|TI|RI
//       0   1   0   0   0   0  0  0
    SCON = 0x40;                         //串行口 8 位 UTRA
    TH1  = 0xFA;
    TL1  = 0xFA;                         //波特率发生器,波特率为 4800
    ET0  = 1;                            //开定时器 T0 中断允许
    ES   = 1;                            //开串行口中断
    EA   = 1;                            //开总中断
    REN  = 1;                            //允许串行接收
    TR1  = 1;                            //开串行口波特率发生器(T1)
    while(1)                             //主程序
      {
        up = 1;                          //把键盘位置 1,以便进行键盘输入
        down = 1;
        keywork();                       //键盘扫描
        if(disp_flag)                    //判断是否可以更新显示
          {
            disp_flag = 0;               //清显示更新标志
            display();                   //显示程序
          }
      }
}
//液晶字模文件(LCD_code.h)
unsigned char code welcome_1[] =
{
/*---------------------------------------
;源文件/文字: Welcome to GPS
;宽×高(像素): 112×16
;字模格式/大小: 单色点阵液晶字模,纵向取模,字节倒序/224 字节
;数据转换日期: 2004 - 5 - 16 20:24:15
---------------------------------------*/
```

0xF8,0x08,0x00,0xF8,0x00,0x08,0xF8,0x00,0x00,0x00,0x80,0x80,0x80,0x80,0x00,0x00,
0x00,0x08,0x08,0xF8,0x00,0x00,0x00,0x00,0x00,0x00,0x80,0x80,0x80,0x00,0x00,0x00,
0x00,0x00,0x80,0x80,0x80,0x80,0x00,0x00,0x80,0x80,0x80,0x80,0x80,0x80,0x80,0x00,
0x00,0x00,0x80,0x80,0x80,0x80,0x00,0x00,0x00,0x00,0x00,0x00,0x00,0x00,0x00,0x00,
0x00,0x80,0x80,0xE0,0x80,0x80,0x00,0x00,0x00,0x80,0x80,0x80,0x80,0x00,0x00,0x00,
0x00,0x00,0x00,0x00,0x00,0x00,0x00,0xC0,0x30,0x08,0x08,0x08,0x38,0x00,0x00,
0x08,0xF8,0x08,0x08,0x08,0x08,0xF0,0x00,0x00,0x70,0x88,0x08,0x08,0x08,0x38,0x00,
0x03,0x3C,0x07,0x00,0x07,0x3C,0x03,0x00,0x00,0x1F,0x22,0x22,0x22,0x22,0x13,0x00,
0x00,0x20,0x20,0x3F,0x20,0x20,0x00,0x00,0x00,0x0E,0x11,0x20,0x20,0x20,0x11,0x00,
0x00,0x1F,0x20,0x20,0x20,0x20,0x1F,0x00,0x20,0x3F,0x20,0x00,0x3F,0x20,0x00,0x3F,
0x00,0x1F,0x22,0x22,0x22,0x22,0x13,0x00,0x00,0x00,0x00,0x00,0x00,0x00,0x00,0x00,
0x00,0x00,0x00,0x1F,0x20,0x20,0x00,0x00,0x00,0x1F,0x20,0x20,0x20,0x20,0x1F,0x00,
0x00,0x00,0x00,0x00,0x00,0x00,0x07,0x18,0x20,0x20,0x22,0x1E,0x02,0x00,
0x20,0x3F,0x21,0x01,0x01,0x01,0x00,0x00,0x00,0x38,0x20,0x21,0x21,0x22,0x1C,0x00
};

unsigned char code welcome_2[] =
{
/*--
;源文件/文字：System
;宽×高(像素)：104×16
;字模格式/大小：单色点阵液晶字模,纵向取模,字节倒序/208字节
;数据转换日期：2004-5-16 20:32:36
--*/
0x00,0x70,0x88,0x08,0x08,0x08,0x38,0x00,0x80,0x80,0x80,0x00,0x00,0x80,0x80,0x80,
0x00,0x00,0x80,0x80,0x80,0x80,0x80,0x00,0x00,0x80,0x80,0xE0,0x80,0x80,0x00,0x00,
0x00,0x00,0x80,0x80,0x80,0x80,0x00,0x00,0x00,0x80,0x80,0x80,0x80,0x80,0x80,0x00,
0x00,0x38,0x20,0x21,0x21,0x22,0x1C,0x00,0x80,0x81,0x8E,0x70,0x18,0x06,0x01,0x00,
0x00,0x33,0x24,0x24,0x24,0x24,0x19,0x00,0x00,0x00,0x00,0x1F,0x20,0x20,0x00,0x00,
0x00,0x1F,0x22,0x22,0x22,0x22,0x13,0x00,0x20,0x3F,0x20,0x00,0x3F,0x20,0x00,0x3F
};

unsigned char code welcome_3[] =
{
/*--
;源文件/文字 ：设计者：
;宽×高(像素)：64×16
;字模格式/大小：单色点阵液晶字模,纵向取模,字节倒序/128字节
;数据转换日期：2004-5-16 20:45:17
--*/
0x40,0x41,0xCE,0x04,0x00,0x80,0x40,0xBE,0x82,0x82,0x82,0xBE,0xC0,0x40,0x40,0x00,
0x20,0x21,0x2E,0xE4,0x00,0x00,0x20,0x20,0x20,0x20,0xFF,0x20,0x20,0x20,0x20,0x00,
0x00,0x20,0x24,0x24,0x24,0x24,0xA4,0xBF,0x64,0x24,0x34,0x28,0x26,0x24,0x20,0x00,
0x00,0x00,0x00,0x00,0x00,0x00,0x00,0x00,0x00,0x00,0x00,0x00,0x00,0x00,0x00,0x00,

```
0x00,0x00,0x7F,0x20,0x90,0x80,0x40,0x43,0x2C,0x10,0x10,0x2C,0x43,0xC0,0x40,0x00,
0x00,0x00,0x00,0x7F,0x20,0x10,0x08,0x00,0x00,0x00,0xFF,0x00,0x00,0x00,0x00,0x00,
0x04,0x04,0x04,0x02,0x02,0xFF,0x49,0x49,0x49,0x49,0x49,0x49,0xFF,0x00,0x00,0x00,
0x00,0x00,0x36,0x36,0x00,0x00,0x00,0x00,0x00,0x00,0x00,0x00,0x00,0x00,0x00,0x00
};

unsigned char code welcome_4[] =
{
/* ----------------------------------------
;源文件/文字：唐召波
;宽×高(像素)：48×16
;字模格式/大小：单色点阵液晶字模,纵向取模,字节倒序/96 字节
;数据转换日期：2004-5-16 20:45:53
   ---------------------------------------- */
0x00,0x00,0xFC,0x44,0x54,0x54,0x54,0x55,0xFE,0x54,0x54,0xF4,0x44,0x44,0x00,0x00,
0x00,0x00,0x82,0x82,0x42,0x22,0x12,0x0E,0x02,0x42,0x42,0xC2,0x42,0x3E,0x00,0x00,
0x10,0x60,0x01,0xE6,0x00,0xF8,0x88,0x88,0x88,0xFF,0x88,0x88,0x88,0x18,0x08,0x00,
0x40,0x30,0x0F,0x00,0x7D,0x25,0x25,0x25,0x27,0x25,0x25,0x7D,0x00,0x00,0x00,0x00,
0x00,0x00,0x00,0x00,0xFF,0x41,0x41,0x41,0x41,0x41,0x41,0x41,0xFF,0x00,0x00,0x00,
0x04,0x04,0xFF,0x40,0x30,0x8F,0x80,0x41,0x26,0x18,0x18,0x26,0x61,0xC0,0x40,0x00
};

unsigned char code jing[] =
{
/* ----------------------------------------
;源文件/文字：经
;宽×高(像素)：16×16
;字模格式/大小：单色点阵液晶字模,纵向取模,字节倒序/32 字节
;数据转换日期：2004-5-16 21:09:10
   ---------------------------------------- */
0x20,0x30,0xAC,0x63,0x10,0x00,0x40,0x42,0x22,0x22,0x12,0x1A,0x26,0x42,0xC0,0x00,
0x22,0x23,0x22,0x12,0x12,0x00,0x20,0x21,0x21,0x21,0x3F,0x21,0x21,0x21,0x20,0x00
};

unsigned char code wei[] =
{
/* ----------------------------------------
;源文件/文字：纬
;宽×高(像素)：16×16
;字模格式/大小：单色点阵液晶字模,纵向取模,字节倒序/32 字节
;数据转换日期：2004-5-16 21:11:14
   ---------------------------------------- */
0x20,0x30,0xAC,0x63,0x20,0x18,0x24,0x24,0x24,0xFF,0x24,0x24,0x24,0x04,0x00,
0x12,0x33,0x12,0x12,0x0A,0x08,0x01,0x01,0x01,0xFF,0x01,0x01,0x11,0x21,0x1F,0x00
};
```

```c
unsigned char code azimuth[4][32] =
{
    {                              //东
    0x00,0x04,0x04,0xC4,0xB4,0x8C,0x87,0x84,0xF4,0x84,0x84,0x84,0x84,0x04,0x00,
    0x00,0x00,0x00,0x20,0x18,0x0E,0x04,0x20,0x40,0xFF,0x00,0x02,0x04,0x18,0x30,
    0x00,0x00
    },
    {                              //南
    0x04,0x04,0xE4,0x24,0x24,0x64,0xB4,0x2F,0x24,0xA4,0x64,0x24,0x24,0xE4,0x04,
    0x00,0x00,0x00,0x7F,0x04,0x05,0x05,0x05,0x7F,0x05,0x05,0x05,0x25,0x44,0x3F,
    0x00,0x00
    },
    {                              //西
    0x02,0xF2,0x12,0x12,0x12,0xFE,0x12,0x12,0x12,0xFE,0x12,0x12,0x12,0xF2,0x02,
    0x00,0x00,0x00,0x7F,0x28,0x24,0x22,0x21,0x20,0x20,0x20,0x21,0x22,0x22,0x22,0x7F,
    0x00,0x00
    },
    {                              //北
    0x00,0x20,0x20,0x20,0x20,0xFF,0x00,0x00,0x00,0xFF,0x40,0x20,0x30,0x18,0x10,
    0x00,0x10,0x30,0x18,0x08,0x04,0x7F,0x00,0x00,0x00,0x3F,0x40,0x40,0x40,0x40,
    0x78,0x00
    },
};
unsigned char code bjsj[] =
{
/*--------------------------------------------
;源文件/文字：北京时间
;宽×高(像素)：64×16
;字模格式/大小：单色点阵液晶字模,纵向取模,字节倒序/128 字节
;数据转换日期：2004-5-16 21:16:11
--------------------------------------------*/
0x00,0x20,0x20,0x20,0x20,0xFF,0x00,0x00,0x00,0xFF,0x40,0x20,0x30,0x18,0x10,0x00,
0x00,0x04,0x04,0xE4,0x24,0x24,0x25,0x26,0x24,0x24,0x24,0xE4,0x04,0x06,0x04,0x00,0x00,
0x00,0xFC,0x44,0x44,0x44,0xFC,0x10,0x90,0x10,0x10,0x10,0xFF,0x10,0x10,0x10,0x00,
0x00,0xF8,0x01,0x06,0x00,0xF0,0x92,0x92,0x92,0x92,0xF2,0x02,0x02,0xFE,0x00,0x00,
0x10,0x30,0x18,0x08,0x04,0x7F,0x00,0x00,0x00,0x3F,0x40,0x40,0x40,0x40,0x78,0x00,
0x00,0x20,0x10,0x19,0x0D,0x41,0x81,0x7F,0x01,0x01,0x05,0x0D,0x38,0x10,0x00,0x00,
0x00,0x07,0x04,0x04,0x04,0x07,0x00,0x00,0x03,0x40,0x80,0x7F,0x00,0x00,0x00,0x00,
0x00,0xFF,0x00,0x00,0x00,0x07,0x04,0x04,0x04,0x04,0x07,0x40,0x80,0x7F,0x00,0x00
};
unsigned char code num[17][16] =
```

```
{
//----------------------------------------
//源文件/文字：0,1,2,3,4,5,6,7,8,9,.,
//宽×高(像素)：8×16
//字模格式/大小：单色点阵液晶字模,纵向取模,字节倒序/128字节
//数据转换日期：2004-5-16 21:16:11
//----------------------------------------
 {0x00,0xE0,0x10,0x08,0x08,0x10,0xE0,0x00,0x00,0x0F,0x10,0x20,0x20,0x10,0x0F,
  0x00}, //0
 {0x00,0x10,0x10,0xF8,0x00,0x00,0x00,0x00,0x00,0x20,0x20,0x3F,0x20,0x20,0x00,
  0x00}, //1
 {0x00,0x70,0x08,0x08,0x08,0x88,0x70,0x00,0x00,0x30,0x28,0x24,0x22,0x21,0x30,
  0x00}, //2
 {0x00,0x30,0x08,0x88,0x88,0x48,0x30,0x00,0x00,0x18,0x20,0x20,0x20,0x11,0x0E,
  0x00}, //3
 {0x00,0x00,0xC0,0x20,0x10,0xF8,0x00,0x00,0x00,0x07,0x04,0x24,0x24,0x3F,0x24,
  0x00}, //4
 {0x00,0xF8,0x08,0x88,0x88,0x08,0x08,0x00,0x00,0x19,0x21,0x20,0x20,0x11,0x0E,
  0x00}, //5
 {0x00,0xE0,0x10,0x88,0x88,0x18,0x00,0x00,0x00,0x0F,0x11,0x20,0x20,0x11,0x0E,
  0x00}, //6
 {0x00,0x38,0x08,0x08,0xC8,0x38,0x08,0x00,0x00,0x00,0x00,0x3F,0x00,0x00,0x00,
  0x00}, //7
 {0x00,0x70,0x88,0x08,0x08,0x88,0x70,0x00,0x00,0x1C,0x22,0x21,0x21,0x22,0x1C,
  0x00}, //8
 {0x00,0xE0,0x10,0x08,0x08,0x10,0xE0,0x00,0x00,0x00,0x31,0x22,0x22,0x11,0x0F,
  0x00}, //9
 {0x00,0x00,0x00,0x00,0x00,0x00,0x00,0x00,0x30,0x30,0x00,0x00,0x00,0x00,
  0x00}, //"."
 {0x00,0x00,0x0C,0x12,0x12,0x0C,0x00,0x00,0x00,0x00,0x00,0x00,0x00,0x00,0x00,
  0x00}, //符号(°)
 {0x10,0x16,0x0E,0x00,0x00,0x00,0x00,0x00,0x00,0x00,0x00,0x00,0x00,0x00,0x00,
  0x00}, //'(分)
 {0x00,0x00,0x00,0xC0,0xC0,0x00,0x00,0x00,0x00,0x00,0x00,0x30,0x30,0x00,0x00,
  0x00}, //":"
 {0x10,0x16,0x0E,0x00,0x10,0x16,0x0E,0x00,0x00,0x00,0x00,0x00,0x00,0x00,0x00,
  0x00}, //""″"(秒)
 {0x08,0xF8,0xF8,0x00,0xF8,0xF8,0x08,0x00,0x20,0x3F,0x00,0x3F,0x00,0x3F,0x20,
  0x00}, //M
 {0x00,0x00,0x00,0x00,0x00,0x00,0x00,0x00,0x00,0x01,0x01,0x01,0x01,0x01,0x01,
  0x01}, //"-"
};
```

```
unsigned char code level[] =
{
/*----------------------------------------
;源文件/文字：大地水准面高度：
;宽×高(像素)：120×16
;字模格式/大小：单色点阵液晶字模，纵向取模，字节倒序/240字节
;数据转换日期：2004-5-21 7:36:49
----------------------------------------*/
0x20,0x20,0x20,0x20,0x20,0x20,0xA0,0x7F,0xA0,0x20,0x20,0x20,0x20,0x20,0x00,
0x10,0x10,0x10,0xFE,0x10,0x50,0x40,0xFE,0x20,0x20,0xFF,0x10,0x10,0xF8,0x10,0x00,
0x00,0x10,0x10,0x10,0x90,0x70,0x00,0xFF,0x20,0x60,0x90,0x08,0x04,0x00,0x00,0x00,
0x00,0x02,0x1C,0xC8,0x20,0x10,0xFF,0x4A,0x48,0x49,0xFE,0x48,0x68,0x4C,0x08,0x00,
0x00,0x02,0xF2,0x12,0x12,0xFA,0x96,0x92,0x92,0xF2,0x12,0x12,0x12,0xF2,0x02,0x00,
0x04,0x04,0x04,0x04,0x74,0x54,0x55,0x56,0x54,0x54,0x74,0x04,0x04,0x04,0x04,0x00,
0x00,0x00,0xFC,0x04,0x24,0x24,0xFC,0xA5,0xA6,0xA4,0xFC,0x24,0x24,0x24,0x04,0x00,
0x00,0x00,0x00,0x00,0xC0,0xC0,0x00,0x00,0x00,0x80,0x40,0x20,0x10,0x0C,0x03,0x00,
0x01,0x06,0x08,0x30,0x60,0xC0,0x40,0x00,0x20,0x20,0x10,0x1F,0x08,0x08,0x00,0x3F,
0x40,0x40,0x4F,0x42,0x44,0x43,0x70,0x00,0x10,0x10,0x08,0x06,0x01,0x40,0x80,0x7F,
0x00,0x00,0x01,0x06,0x0C,0x18,0x08,0x00,0x02,0x02,0x7E,0x01,0x00,0x00,0x7F,0x22,
0x22,0x22,0x3F,0x22,0x23,0x32,0x20,0x00,0x00,0x7F,0x20,0x20,0x3F,0x24,0x24,
0x24,0x3F,0x20,0x20,0x20,0x7F,0x00,0x00,0x00,0xFF,0x01,0x01,0x3D,0x25,0x25,0x25,
0x25,0x25,0x3D,0x41,0x81,0x7F,0x00,0x00,0x80,0x60,0x1F,0x80,0x80,0x42,0x46,0x2A,
0x12,0x12,0x2A,0x26,0x42,0xC0,0x40,0x00,0x00,0x00,0x30,0x30,0x00,0x00,0x00
};

unsigned char code nosignal[] =
{
/*----------------------------------------
;源文件/文字：无GPS接收信号！
;宽×高(像素)：112×16
;字模格式/大小：单色点阵液晶字模，纵向取模，字节倒序/224字节
;数据转换日期：2004-5-21 7:48:48
----------------------------------------*/
0x00,0x40,0x42,0x42,0x42,0x42,0x42,0xFE,0xC2,0x42,0x42,0x42,0x42,0x42,0x40,0x00,
0xC0,0x30,0x08,0x08,0x08,0x38,0x00,0x00,0x08,0xF8,0x08,0x08,0x08,0x08,0xF0,0x00,
0x00,0x70,0x88,0x08,0x08,0x08,0x38,0x00,0x08,0x08,0x08,0xFF,0x88,0x68,0x24,0x2C,
0xB4,0x25,0x26,0x34,0x2C,0x24,0x20,0x00,0x00,0xF8,0x00,0x00,0xFF,0x00,0x40,0x30,
0xDF,0x10,0x10,0x10,0xF0,0x10,0x10,0x00,0x80,0x40,0x30,0xFC,0x07,0x0A,0xA8,0xA8,
0xA9,0xAE,0xAA,0xA8,0xA8,0x08,0x08,0x00,0x40,0x40,0x40,0x5F,0xD1,0x51,0x51,0x51,
0x51,0x51,0x51,0x5F,0x40,0x40,0x40,0x00,0x00,0x00,0xF8,0x00,0x00,0x00,0x00,
0x40,0x40,0x20,0x10,0x08,0x04,0x03,0x00,0x3F,0x40,0x40,0x40,0x40,0x40,0x70,0x00,
0x07,0x18,0x20,0x20,0x22,0x1E,0x02,0x00,0x20,0x3F,0x21,0x01,0x01,0x01,0x00,0x00,
0x00,0x38,0x20,0x21,0x21,0x22,0x1C,0x00,0x02,0x42,0x81,0x7F,0x02,0x82,0x8A,0x4E,
```

第 12 章 液晶 GPS 定位信息显示器的设计

0x53,0x32,0x12,0x2E,0x42,0xC2,0x02,0x00,0x00,0x0F,0x04,0x02,0xFF,0x40,0x40,0x20,
0x21,0x16,0x08,0x16,0x21,0x60,0x20,0x00,0x00,0x00,0x00,0x7F,0x00,0x00,0x7E,0x22,
0x22,0x22,0x22,0x22,0x7E,0x00,0x00,0x00,0x00,0x00,0x02,0x07,0x02,0x02,0x22,
0x42,0x82,0x42,0x3E,0x00,0x00,0x00,0x00,0x00,0x00,0x33,0x30,0x00,0x00,0x00
};

附录 A 80C51 系列单片机的特殊功能寄存器表

80C51 系列单片机的特殊功能寄存器表如表 A.1 所列。

表 A.1 80C51 系列单片机的特殊功能寄存器表

符号	寄存器名	位地址、位标记及位功能								直接地址 addrect	复位状态
		D7	D6	D5	D4	D3	D2	D1	D0		
(1) 可位寻址 SFR(共 11 个)											
ACC	累加器	E7	E6	E5	E4	E3	E2	E1	E0	E0H	00H
		ACC.7	ACC.6	ACC.5	ACC.4	ACC.3	ACC.2	ACC.1	ACC.0		
B	B 寄存器	F7	F6	F5	F4	F3	F2	F1	F0	F0H	00H
		B.7	B.6	B.5	B.4	B.3	B.2	B.1	B.0		
PSW	程序状态字	D7	D6	D5	D4	D3	D2	D1	D0	D0H	00H
		CY	AC	F0	RS1	RS0	OV	—	P		
IP	中断优先权寄存器	BF	BE	BD	BC	BB	BA	B9	B8	B8H	×××00000B
		—	—	—	PS	PT1	PX1	PT0	PX0		
P3	P3 口	B7	B6	B5	B4	B3	B2	B1	B0	B0H	FFH
		P3.7	P3.6	P3.5	P3.4	P3.3	P3.2	P3.1	P3.0		
IE	中断允许寄存器	AF	AE	AD	AC	AB	AA	A9	A8	A8H	0××00000B
		EA	—	—	ES	ET1	EX1	ET0	EX0		
P2	P2 口	A7	A6	A5	A4	A3	A2	A1	A0	A0H	FFH
		P2.7	P2.6	P2.5	P2.4	P2.3	P2.2	P2.1	P2.0		
SCON	串行口控制寄存器	9F	9E	9D	9C	9B	9A	99	98	98H	00H
		SM0	SM1	SM2	REN	TB8	RB8	TI	RI		
P1	P1 口	97	96	95	94	93	92	91	90	90H	FFH
		P1.7	P1.6	P1.5	P1.4	P1.3	P1.2	P1.1	P1.0		
TCON	定时器控制寄存器	8F	8E	8D	8C	8B	8A	89	88	88H	00H
		TF1	TR1	TF0	TR0	IE1	IT1	IE0	IT0		
P0	P0 口	87	86	85	84	83	82	81	80	80H	FFH
		P0.7	P0.6	P0.5	P0.4	P0.3	P0.2	P0.1	P0.0		

附录 A 80C51 系列单片机的特殊功能寄存器表

续表 A.1

符 号	寄存器名	位地址、位标记及位功能								直接地址 addrect	复位状态
		D7	D6	D5	D4	D3	D2	D1	D0		
(2) 不可位寻址 SFR(共 10 个)											
SP	栈指示器									81H	07H
DPL	数据指针低 8 位									82H	00H
DPH	数据指针高 8 位									83H	00H
PCON	电源控制寄存器	SMOD	—	—	—	GF1	GF0	PD	IDL	87H	0×××0000B
TMOD	定时器方式寄存器	GATE	C/$\bar{T}$	M1	M0	GATE	C/$\bar{T}$	M1	M0	89H	00H
TL0	T0 寄存器低 8 位									8AH	00H
TL1	T1 寄存器低 8 位									8BH	00H
TH0	T0 寄存器高 8 位									8CH	00H
TH1	T1 寄存器高 8 位									8DH	00H
SBUF	串行口数据缓冲器									99H	×××× ××××B

附录 B 80C51 系列单片机中断入口地址表

80C51 系列单片机中断入口地址表如表 B.1 所列。

表 B.1 80C51 系列单片机中断入口地址表

ROM 地址	用 途	优先级
0000H	复位程序运行入口地址	
0003H	外中断 0 入口地址	高
000BH	定时器 T0 溢出中断入口地址	↓
0013H	外中断 1 入口地址	
001BH	定时器 T1 溢出中断入口地址	
0023H	串行口发送/接收中断入口地址	低
002BH	定时器 T2 溢出中断入口地址	

附录 C　80C51 系列单片机汇编指令表

80C51 系列单片机汇编指令表如表 C.1～表 C.5 所列。

1. 数据传送指令(29 条)

表 C.1　数据传送指令

汇编指令		操作说明	代码长度/字节	指令周期 T_{osc}	指令周期 T_m
(1) 程序存储器查表指令(共 2 条)					
MOVC	A,@A+DPTR	将以 DPTR 为基址，A 为偏移地址中的数送入 A 中	1	24	2
MOVC	A,@A+PC	将以 PC 为基址，A 为偏移地址中的数送入 A 中	1	24	2
(2) 片外 RAM 传送指令(共 4 条)					
MOVX	A,@DPTR	将片外 RAM 中的 DPTR 地址中的数送入 A 中	1	24	2
MOVX	@DPTR,A	将 A 中的数送入片外 RAM 中的 DPTR 地址单元中	1	24	2
MOVX	A,@Ri	将片外 RAM 中@Ri 指示的地址中的数送入 A 中	1	24	2
MOVX	@Ri,A	将 A 中的数送入片外@Ri 指示的地址单元中	1	24	2
(3) 片内 RAM 及寄存器间数据传送指令(共 18 条)					
MOV	A,Rn	将 Rn 中的数送入 A 中	1	12	1
MOV	A,direct	将直接地址 direct 中的数送入 A 中	2	12	1
MOV	A,#data	将 8 位常数送入 A 中	2	12	1
MOV	A,@Ri	将 Ri 指示的地址中的数送入 A 中	1	12	1
MOV	Rn,direct	将直接地址 direct 中的数送入 Rn 中	2	24	2
MOV	Rn,#data	将立即数送入 Rn 中	2	12	1
MOV	Rn,A	将 A 中的数送入 Rn 中	1	12	1
MOV	direct,Rn	将 Rn 中的数送入 direct 中	2	24	2
MOV	direct,A	将 A 中的数送入 direct 中	2	12	1
MOV	direct,@Ri	将@Ri 指示单元中的数送入 direct 中	2	24	2
MOV	direct,#data	将立即数送入 direct 中	3	24	2

续表 C.1

汇编指令		操作说明	代码长度/字节	指令周期	
				T_{osc}	T_m
MOV	direct,direct	将一个 direct 中的数送入另一个 direct 中	3	24	2
MOV	@Ri,A	将 A 中的数送入 Ri 指示的地址中	1	12	1
MOV	@Ri,direct	将 direct 中的数送入 Ri 指示的地址中	2	24	2
MOV	@Ri,#data	将立即数送入 Ri 指示的地址中	2	12	1
MOV	DPTR,#data16	将 16 位立即数直接送入 DPTR 中	3	24	2
PUSH	direct	将 direct 中的数压入堆栈	2	24	2
POP	direct	将堆栈中的数弹出到 direct 中	2	24	2
(4) 数据交换指令(共 5 条)					
XCH	A,Rn	A 中的数和 Rn 中的数全交换	1	12	1
XCH	A,direct	A 中的数和 direct 中的数全交换	2	12	1
XCH	A,@Ri	A 中的数和 @Ri 中的数全交换	1	12	1
XCHD	A,@Ri	A 中的数和 @Ri 中的数半交换	1	12	1
SWAP	A	A 中的数自交换(高 4 位与低 4 位)	1	12	1

2. 算术运算类指令(共 24 条)

表 C.2 算术运算类指令

汇编指令		操作说明	代码长度/字节	指令周期	
				T_{osc}	T_m
ADD	A,Rn	Rn 中与 A 中的数相加,结果在 A 中,影响 PSW 位的状态	1	12	1
ADD	A,direct	direct 中与 A 中的数相加,结果在 A 中,影响 PSW 位的状态	2	12	1
ADD	A,#data	立即数与 A 中的数相加,结果在 A 中,影响 PSW 位的状态	2	12	1
ADD	A,@Ri	@Ri 中与 A 中的数相加,结果在 A 中,影响 PSW 位的状态	1	12	1
ADDC	A,Rn	Rn 中与 A 中的数带进位加,结果在 A 中,影响 PSW 位的状态	1	12	1
ADDC	A,direct	direct 中与 A 中的数带进位加,结果在 A 中,影响 PSW 位的状态	2	12	1

续表 C.2

汇编指令		操作说明	代码长度/字节	指令周期	
				T_{osc}	T_m
ADDC	A,#data	立即数与 A 中的数带进位加,结果在 A 中,影响 PSW 位的状态	2	12	1
ADDC	A,@Ri	@Ri 中与 A 中的数带进位加,结果在 A 中,影响 PSW 位的状态	1	12	1
SUBB	A,Rn	Rn 中与 A 中的数带借位减,结果在 A 中,影响 PSW 位的状态	1	12	1
SUBB	A,direct	direct 中与 A 中的数带借位减,结果在 A 中,影响 PSW 位的状态	2	12	1
SUBB	A,#data	立即数与 A 中的数带借位减,结果在 A 中,影响 PSW 位的状态	2	12	1
SUBB	A,@Ri	@Ri 中与 A 中的数带借位减,结果在 A 中,影响 PSW 位的状态	1	12	1
INC	A	A 中数加 1	1	12	1
INC	Rn	Rn 中数加 1	1	12	1
INC	direct	Direct 中数加 1	2	12	1
INC	@Ri	@Ri 中数加 1	1	12	1
INC	DPTR	DPTR 中数加 1	1	24	2
DEC	A	A 中数减 1	1	12	1
DEC	Rn	Rn 中数减 1	1	12	1
DEC	direct	Direct 中数减 1	2	12	1
DEC	@Ri	@Ri 中数减 1	1	12	1
MUL	AB	A,B 中两无符号数相乘,结果低 8 位在 A 中,高 8 位在 B 中	1	48	4
DIV	AB	A,B 中两无符号数相除,商在 A 中,余数在 B 中	1	48	4
DA	A	十进制调整,对 BCD 码十进制加法运算结果调整	1	12	1

3. 逻辑运算指令(共 24 条)

表 C.3　逻辑运算指令

汇编指令		操作说明	代码长度/字节	指令周期	
				T_{osc}	T_m
ANL	A,Rn	Rn 中与 A 中的数相"与",结果在 A 中	1	12	1
ANL	A,direct	direct 中与 A 中的数相"与",结果在 A 中	2	12	1
ANL	A,#data	立即数与 A 中的数相"与",结果在 A 中	2	12	1
ANL	A,@Ri	@Ri 中与 A 中的数相"与",结果在 A 中	1	12	1
ANL	direct,A	A 和 direct 中的数进行"与"操作,结果在 direct 中	2	12	1
ANL	direct,#data	常数和 direct 中的数进行"与"操作,结果在 direct 中	3	24	2
ORL	A,Rn	Rn 中和 A 中的数进行"或"操作,结果在 A 中	1	12	1
ORL	A,direct	direct 中和 A 中的数进行"或"操作,结果在 A 中	2	12	1
ORL	A,#data	立即数和 A 中的数进行"或"操作,结果在 A 中	2	12	1
ORL	A,@Ri	@Ri 中和 A 中的数进行"或"操作,结果在 A 中	1	12	1
ORL	direct,A	A 中和 direct 中的数进行"或"操作,结果在 direct 中	2	12	1
ORL	direct,#data	立即数和 direct 中的数进行"或"操作,结果在 direct 中	3	24	2
XRL	A,Rn	Rn 中和 A 中的数进行"异或"操作,结果在 A 中	1	12	1
XRL	A,direct	direct 中和 A 中的数进行"异或"操作,结果在 A 中	2	12	1
XRL	A,#data	立即数和 A 中的数进行"异或"操作,结果在 A 中	2	12	1
XRL	A,@Ri	@Ri 中和 A 中的数进行"异或"操作,结果在 A 中	1	12	1
XRL	direct,A	A 中和 direct 中的数进行"异或"操作,结果在 direct 中	2	12	1
XRL	direct,#data	立即数和 direct 中的数进行"异或"操作,结果在 direct 中	3	24	2
RR	A	A 中的数循环右移(移向低位),D0 移入 D7	1	12	1
RRC	A	A 中的数带进位循环右移,D0 移入 C,C 移入 D7	1	12	1
RL	A	A 中的数循环左移(移向高位),D7 移入 D0	1	12	1
RLC	A	A 中的数带进位循环左移,D7 移入 C,C 移入 D0	1	12	1
CLR	A	A 中的数清 0	1	12	1
CPL	A	A 中的数取反	1	12	1

4. 程序转移类指令(共17条)

表C.4 程序转移类指令

汇编指令		操作说明	代码长度/字节	指令周期	
				T_{osc}	T_m
(1) 无条件转移指令(共9条)					
LJMP	addr16	长转移,程序转到addr16指示的地址处	3	24	2
AJMP	addr11	短转移,程序转到addr11指示的地址处	2	24	2
SJMP	rel	相对转移,程序转到rel指示的地址处	2	24	2
LCALL	addr16	长调用,程序调用addr16处的子程序	3	24	2
ACALL	addr11	短调用,程序调用addr11处的子程序	2	24	2
JMP	@A+DPTR	程序散转,程序转到DPTR为基址,A为偏移地址处	1	24	2
RETI		中断返回	1	24	2
RET		子程序返回	1	24	2
NOP		空操作	1	12	1
(2) 条件转移指令(共8条)					
JZ	rel	A中的数为0,程序转到相对地址rel处	2	24	2
JNZ	rel	A中的数不为0,程序转到相对地址rel处	2	24	2
DJNZ	Rn,rel	Rn中的数减1不为0,程序转到相对地址rel处	2	24	2
DJNZ	direct,rel	direct中的数减1不为0,程序转到相对地址rel处	3	24	2
CJNE	A,#data,rel	#data与A中的数不等,转至rel处。C=1,data>(A);C=0,data<(A)	3	24	2
CJNE	A,direct,rel	direct与A中的数不等,转至rel处。C=1,data>(A);C=0,data<(A)	3	24	2
CJNE	Rn,#data,rel	#data与Rn中的数不等,转至rel处。C=1,data>(Rn);C=0,data<(Rn)	3	24	2
CJNE	@Ri,#data,rel	#data与@Ri中的数不等,转至rel处。C=1,data>(@Ri);C=0,data<(@Ri)	3	24	2

5. 布尔指令(共 17 条)

表 C.5 布尔指令

汇编指令		操作说明	代码长度/字节	指令周期 T_{osc}	指令周期 T_m
(1) 位操作指令(共 12 条)					
MOV	C,bit	bit 中的状态送入 C 中	2	12	1
MOV	bit,C	C 中的状态送入 bit 中	2	24	2
ANL	C,bit	bit 中的状态与 C 中的状态相"与",结果在 C 中	2	24	2
ANL	C,/bit	bit 中的状态取反与 C 中的状态相"与",结果在 C 中	2	24	2
ORL	C,bit	bit 中的状态与 C 中的状态相"或",结果在 C 中	2	24	2
ORL	C,/bit	bit 中的状态取反与 C 中的状态相"或",结果在 C 中	2	24	2
CLR	C	C 中的状态清 0	1	12	1
SETB	C	C 中的状态置 1	1	12	1
CPL	C	C 中的状态取反	1	12	1
CLR	bit	bit 中的状态清 0	2	12	1
SETB	bit	bit 中的状态置 1	2	12	1
CPL	bit	bit 中的状态取反	2	12	1
(2) 位条件转移指令(共 5 条)					
JC	rel	进位位为 0 时,程序转至 rel	2	24	2
JNC	rel	进位位不为 0 时,程序转至 rel	2	24	2
JB	bit,rel	bit 状态为 1 时,程序转至 rel	3	24	2
JNB	bit,rel	bit 状态不为 1 时,程序转至 rel	3	24	2
JBC	bit,rel	bit 状态为 1 时,程序转至 rel,同时 bit 位清 0	3	24	2

参考文献

[1] 楼然苗,李光飞. 单片机课程设计指导[M]. 2版. 北京:北京航空航天大学出版社,2012.

[2] 楼然苗,胡佳文,李光飞,等. 单片机实验与课程设计指导(Proteus仿真版)[M]. 2版. 杭州:浙江大学出版社,2013.

[6] 楼然苗,胡佳文,李光飞,等. 51系列单片机原理及应用[M]. 北京:北京航空航天大学出版社,2014.